Content-based Retrieval of Medical Images

Landmarking, Indexing, and Relevance Feedback

Synthesis Lectures on Biomedical Engineering

Editor
John D. Enderle, *University of Connecticut*

Lectures in Biomedical Engineering will be comprised of 75- to 150-page publications on advanced and state-of-the-art topics that span the field of biomedical engineering, from the atom and molecule to large diagnostic equipment. Each lecture covers, for that topic, the fundamental principles in a unified manner, develops underlying concepts needed for sequential material, and progresses to more advanced topics. Computer software and multimedia, when appropriate and available, are included for simulation, computation, visualization and design. The authors selected to write the lectures are leading experts on the subject who have extensive background in theory, application and design.

The series is designed to meet the demands of the 21st century technology and the rapid advancements in the all-encompassing field of biomedical engineering that includes biochemical processes, biomaterials, biomechanics, bioinstrumentation, physiological modeling, biosignal processing, bioinformatics, biocomplexity, medical and molecular imaging, rehabilitation engineering, biomimetic nano-electrokinetics, biosensors, biotechnology, clinical engineering, biomedical devices, drug discovery and delivery systems, tissue engineering, proteomics, functional genomics, and molecular and cellular engineering.

Content-based Retrieval of Medical Images: Landmarking, Indexing, and Relevance Feedback
Paulo Mazzoncini de Azevedo-Marques and Rangaraj Mandayam Rangayyan
2013

Computer-aided Detection of Architectural Distortion in Prior Mammograms of Interval Cancer
Shantanu Banik, Rangaraj M. Rangayyan, and J.E. Leo Desautels
2013

Chronobioengineering: Introduction to Biological Rhythms with Applications, Volume 1
Donald McEachron
2012

The Graph Theoretical Approach in Brain Functional Networks: Theory and Applications
Fabrizio De Vico Fallani and Fabio Babiloni
2010

Biomedical Technology Assessment: The 3Q Method
Phillip Weinfurt
2010

Strategic Health Technology Incorporation
Binseng Wang
2009

Phonocardiography Signal Processing
Abbas K. Abbas and Rasha Bassam
2009

Introduction to Biomedical Engineering: Biomechanics and Bioelectricity - Part II
Douglas A. Christensen
2009

Introduction to Biomedical Engineering: Biomechanics and Bioelectricity - Part I
Douglas A. Christensen
2009

Landmarking and Segmentation of 3D CT Images
Shantanu Banik, Rangaraj M. Rangayyan, and Graham S. Boag
2009

Basic Feedback Controls in Biomedicine
Charles S. Lessard
2009

Understanding Atrial Fibrillation: The Signal Processing Contribution, Part I
Luca Mainardi, Leif Sörnmo, and Sergio Cerutti
2008

Understanding Atrial Fibrillation: The Signal Processing Contribution, Part II
Luca Mainardi, Leif Sörnmo, and Sergio Cerutti
2008

Introductory Medical Imaging
A. A. Bharath
2008

Lung Sounds: An Advanced Signal Processing Perspective
Leontios J. Hadjileontiadis
2008

BioNanotechnology
Elisabeth S. Papazoglou and Aravind Parthasarathy
2007

Bioinstrumentation
John D. Enderle
2006

Fundamentals of Respiratory Sounds and Analysis
Zahra Moussavi
2006

Advanced Probability Theory for Biomedical Engineers
John D. Enderle, David C. Farden, and Daniel J. Krause
2006

Intermediate Probability Theory for Biomedical Engineers
John D. Enderle, David C. Farden, and Daniel J. Krause
2006

Basic Probability Theory for Biomedical Engineers
John D. Enderle, David C. Farden, and Daniel J. Krause
2006

Sensory Organ Replacement and Repair
Gerald E. Miller
2006

Artificial Organs
Gerald E. Miller
2006

Signal Processing of Random Physiological Signals
Charles S. Lessard
2006

Image and Signal Processing for Networked E-Health Applications
Ilias G. Maglogiannis, Kostas Karpouzis, and Manolis Wallace
2006

Content-based Retrieval of Medical Images: Landmarking, Indexing, and Relevance Feedback

Paulo Mazzoncini de Azevedo-Marques and Rangaraj Mandayam Rangayyan

ISBN: 978-3-031-00523-7   paperback
ISBN: 978-3-031-01651-6   ebook

DOI: 10.1007/978-3-031-01651-6

A Publication in the Springer series
SYNTHESIS LECTURES ON BIOMEDICAL ENGINEERING

Lecture #48
Series Editor: John D. Enderle, *University of Connecticut*
Series ISSN
Synthesis Lectures on Biomedical Engineering
Print 1930-0328   Electronic 1930-0336

Content-based Retrieval of Medical Images

Landmarking, Indexing, and Relevance Feedback

Paulo Mazzoncini de Azevedo-Marques
University of São Paulo, Ribeirão Preto, São Paulo, Brazil

Rangaraj Mandayam Rangayyan
University of Calgary, Calgary, Alberta, Canada

SYNTHESIS LECTURES ON BIOMEDICAL ENGINEERING #48

ABSTRACT

Content-based image retrieval (CBIR) is the process of retrieval of images from a database that are similar to a query image, using measures derived from the images themselves, rather than relying on accompanying text or annotation. To achieve CBIR, the contents of the images need to be characterized by quantitative features; the features of the query image are compared with the features of each image in the database and images having high similarity with respect to the query image are retrieved and displayed. CBIR of medical images is a useful tool and could provide radiologists with assistance in the form of a display of relevant past cases. One of the challenging aspects of CBIR is to extract features from the images to represent their visual, diagnostic, or application-specific information content.

In this book, methods are presented for preprocessing, segmentation, landmarking, feature extraction, and indexing of mammograms for CBIR. The preprocessing steps include anisotropic diffusion and the Wiener filter to remove noise and perform image enhancement. Techniques are described for segmentation of the breast and fibroglandular disk, including maximum entropy, a moment-preserving method, and Otsu's method. Image processing techniques are described for automatic detection of the nipple and the edge of the pectoral muscle via analysis in the Radon domain. By using the nipple and the pectoral muscle as landmarks, mammograms are divided into their internal, external, upper, and lower parts for further analysis. Methods are presented for feature extraction using texture analysis, shape analysis, granulometric analysis, moments, and statistical measures.

The CBIR system presented provides options for retrieval using the Kohonen self-organizing map and the k-nearest-neighbor method. Methods are described for inclusion of expert knowledge to reduce the semantic gap in CBIR, including the query point movement method for relevance feedback (RFb). Analysis of performance is described in terms of precision, recall, and relevance-weighted precision of retrieval. Results of application to a clinical database of mammograms are presented, including the input of expert radiologists into the CBIR and RFb processes. Models are presented for integration of CBIR and computer-aided diagnosis (CAD) with a picture archival and communication system (PACS) for efficient workflow in a hospital.

KEYWORDS

anisotropic diffusion, breast cancer, breast density, computer-aided diagnosis, content-based image retrieval, fibroglandular disk, granulometry, image enhancement, image segmentation, information retrieval, Kohonen self-organizing map, landmarking of images, mammography, nipple detection, pattern recognition, pectoral muscle, picture archival and communication system, Radon transform, relevance feedback, texture analysis, Wiener filter

Paulo M. Azevedo-Marques dedicates this book to his great aunt Elina Barbucci, wife Maria Eunice M. Azevedo-Marques, and children Giulia M. Azevedo-Marques and Luca M. Azevedo-Marques

Contents

Preface

The aim of this book is to present some of the recent developments in the areas of CBIR and CAD, with particular reference to mammography and breast cancer. In spite of several advances made in medical imaging and image analysis, there are practical difficulties in incorporating new methods and systems into the workflow in a hospital or clinical environment. With the introduction of PACS, efficient ways are available to introduce new computational models and procedures into radiological and clinical practice. The relatively new concept of RFb facilitates not only the incorporation of the expertise and preferences of the users but also improved training of the CBIR system. In this book, we explain in detail the various procedures involved in the concepts mentioned above and demonstrate practical applications. The results presented in the book provide evidence of success and indicate potential for further improvement.

The general concepts of CBIR and CAD presented in this book may be easily understood by any reader with a university-level education in any area. In particular, the models and general methodologies may be appreciated readily by medical specialists such as radiologists as well as affiliated specialists including clinicians and medical physicists. The mathematical procedures and algorithms presented require background in signal or image processing, and may be used by engineers and computer scientists to implement related procedures.

We wish the reader an intellectually challenging experience with the new concepts presented in the book as well as success in realizing them in practice.

Paulo Mazzoncini de Azevedo-Marques
Rangaraj Mandayam Rangayyan
January 2013

Acknowledgments

We thank the radiologists and faculty members of the Clinical Hospital of the Faculty of Medicine, University of São Paulo, Ribeirão Preto, SP, Brazil, for providing the mammograms and the related reports used in this work, and for testing the system. We thank the State of São Paulo Research Foundation (FAPESP), the National Council for Scientific and Technological Development (CNPq) of Brazil, and the Foundation to Aid Teaching, Research, and Patient Care of the Clinical Hospital of Ribeirão Preto (FAEPA/HCRP) for financial support. This work was also supported by a Catalyst grant from Research Services of the University of Calgary, Calgary, AB, Canada.

We thank our collaborators and students who have contributed to the research work underlying the present book: Dr. Natália Abdala Rosa, Dr. Sérgio Koodi Kinoshita, Dr. Agma Juci Machado Traina, Dr. Caetano Traina Jr., Dr. Roberto Rodrigues Pereira Jr., Dr. José Antônio Heisinger Rodrigues, and Dr. Marcello Henrique Nogueira Barbosa.

Paulo Mazzoncini de Azevedo-Marques
Rangaraj Mandayam Rangayyan
January 2013

Symbols and Abbreviations

arg	argument of
ACR	American College of Radiology
ADO	ActiveX data objects
ASP	active server pages
A_z	area under the ROC curve
b	bit
bpp	bits per pixel
B	byte
BI-RADS^{TM}	Breast Imaging–Reporting and Data System
bps	bits per second
cbPACS	content-based PACS
cm	centimeter
C	compactness
CAD	computer-aided diagnosis
CBIR	content-based image retrieval
CC	craniocaudal
CR	computed radiography
CT	computed tomography
CTN	central test node
div	divergence
dpi	dots per inch
DBMS	database management system
DICOM	Digital Imaging and Communications in Medicine
DICOM-SR	DICOM structured report
DIMSE	DICOM message service element
DR	digital radiography
DR	ratio of diameters
E[ ]	statistical expectation
EHR	electronic health record
$f(m, n)$	a digital image, typically original or undistorted
$f(x, y)$	an image, typically original or undistorted
FN	false negative

FNF	false-negative fraction
FP	false positive
FPF	false-positive fraction
$g(m, n)$	a digital image, typically processed or distorted
$g(x, y)$	an image, typically processed or distorted
H	entropy
HCFMRP	Hospital das Clínicas da Faculdade de Medicina de Ribeirão Preto
	Medical Center of the School of Medicine of Ribeirão Preto
HTML	Hyper Text Markup Language
$\mathbf{I}$	identity matrix
IHE	integrating the healthcare enterprise
IIS	internet information server
ISO	International Standards Organization
ISO-OSI	ISO open systems interconnection
IT	information technology
j	$\sqrt{-1}$
JPEG	Joint Photographic Experts Group
k	kilo (1,000)
k-NN	k-nearest neighbors
kVp	kilo-volt peak
K	kilo (1,024)
ln	natural log (to base e)
LLMMSE	local linear minimum mean-squared error
LMMSE	linear minimum mean-squared error
LMS	least mean squares
m	meter
m	moment
max	maximum
Mbps	megabits per second
min	minimum
mm	millimeter
(m, n)	indices in the discrete space (image) domain
M	mega
M	moment
M	number of samples or pixels
MIAS	Mammographic Image Analysis Society, London, UK
MLO	mediolateral oblique
MMSE	minimum mean-squared error
MPP	(relevance feedback using) multiple point projection

MR	magnetic resonance
MRI	magnetic resonance imaging
MS	mean-squared
MSE	mean-squared error
n	an index
N	number of samples or pixels
NEMA	National Electrical Manufacturers Association
PET	positron emission tomography
pixel	picture cell or element
$p(x)$	probability density function of the random variable x
$P_f(l)$	histogram of image f
$P(x)$	probability of the event x
PACS	picture archival and communication system
PCA	principal component analysis
PDF	probability density function
QPM	query point movement
RADR	(diagnostic report by) radiologist "R"
RFb	relevance feedback
RFP	relevance feedback projection
RIS	radiology information system
ROC	receiver operating characteristics
ROI	region of interest
RSNA	Radiological Society of North America
RWP	relevance-weighted precision
$R(t, \theta)$	projection (Radon transform) of an image at angle θ
s	space variable in the projection (Radon) space
SC	secondary capture
SCP	DICOM service class provider
SCU	DICOM service class user
SNR	signal-to-noise ratio
SOM	self-organizing map
SOP	DICOM service object pair
SPECT	single-photon emission computed tomography
SVM	support vector machine
t	space variable in the projection (Radon) space
T	a threshold
T	as a superscript, vector or matrix transposition
voxel	volume cell or element
V	volt

(x, y)	image coordinates in the space domain
1D	one-dimensional
2D	two-dimensional
3D	three-dimensional
4D	four-dimensional
ε, ϵ	model error, total squared error
δ	Dirac delta function
η	a random variable or noise process
θ	an angle
(t, θ)	the Radon (projection) space
μ	X-ray attenuation coefficient
μ	mean
μ	moment
μm	micrometer
σ	the standard deviation of a random variable
σ^2	the variance of a random variable
σ_{fg}	covariance between images f and g
∂	partial derivative
∇	gradient operator
$*$	when in-line, convolution
$*$	as a superscript, complex conjugation
$-$	average or normalized version of the variable under the bar
$\sim$	estimate of the variable under the symbol
$\forall$	for all
$\in$	belongs to or is in (the set)
$\{\}$	a set
$\subset$	subset
$\supset$	superset
$\cap$	intersection
$\cup$	union
$\equiv$	equivalent to
$\vert$	given, conditional upon
$[\,]$	closed interval, including the limits
$(\,)$	open interval, not including the limits
$\vert\,\vert$	absolute value or magnitude
$\vert\,\vert$	determinant of a matrix
$\Vert\ \Vert$	norm of a vector or matrix
$\lceil x \rceil$	ceiling operator; the smallest integer $\geq x$
$\lfloor x \rfloor$	floor operator; the largest integer $\leq x$

CHAPTER 1

Introduction to Content-based Image Retrieval

1.1 DIGITAL IMAGING IN MEDICAL DIAGNOSTICS

The field of medical imaging has been well established in medical diagnostics since the discovery of X rays by Röntgen in 1895. The development of practical computed tomography (CT) scanners in the early 1970s by Hounsfield and others brought computers and digital images into medical imaging and clinical practice in radiology. Since then, computers and digital imaging have been integrated into radiology and medical imaging departments in hospitals. Computers are now routinely used to perform a variety of tasks from data acquisition and image generation to image display and analysis [1–4].

With the development of new imaging modalities, the need for computers and computing in image generation, manipulation, display, visualization, and analysis continued to grow. Computers are now a part of almost every medical imaging system, including radiography, ultrasonography, CT, nuclear medicine, and magnetic resonance (MR) imaging (MRI) systems. Most radiology departments have turned themselves into "totally digital" and "filmless" departments, using computers for image archival and communication as well, via the establishment of picture archival and communication systems (PACS). The X-ray film that launched the field of radiological imaging has almost vanished. Most of the diagnostic work is now performed using computers not only for the display of image data but also to derive measures from the images and to analyze them: this has led to the establishment of a new field known as computer-aided diagnosis (CAD) [1–3].

With increasing numbers of modalities of medical imaging and their expanding use in routine clinical work, there has been a natural increase in the scope and complexity of the associated problems, calling for further advanced techniques for their solution. For example, X-ray imaging with low-dose digital detectors gives rise to substantial amounts of noise in the resulting images; furthermore, the contrast and visibility of the details in the images may be well below the levels required for efficient detection of abnormalities by human observers. With widespread acceptance of mammography as a screening tool, there is the need to process efficiently such images using techniques of computer vision. The use of high-resolution imaging devices in digital mammography and digital radiography as well as the widespread adoption of PACS in hospitals have created the need for higher levels of data compression and database management systems (DBMS). It is now common to use multiple modalities of medical imaging for improved diagnosis of a particular type of disease; this has raised the need to combine diverse images of the same organ or patient, or the results of their processing, into easily comprehensible multimodal displays.

The development of appropriate techniques for CAD, PACS, and other tools for use in clinical routine requires detailed and thorough understanding of the nature of the images being used, their characteristics and properties, and the needs as well as the preferences of the users. The following sections provide discussions on related topics and issues.

1.2   ANATOMICAL AND PHYSIOLOGICAL FEATURES IN MEDICAL IMAGES

The various modalities of medical imaging mentioned in the preceding section provide widely varying views of parts of the human body [3–6]. In the traditional form of X-ray imaging, the three-dimensional (3D) body is projected onto a two-dimensional (2D) plane. In this mode of imaging, the net effect visualized at each pixel in the image represents the integrated effect of attenuation of the X rays by all of the tissues in the 3D body along each path or ray from the X-ray source to the detector (see Section 2.2.1). Given that attenuation is related to density, planar radiography is useful in discriminating between soft tissue and hard tissue. In radiography, a 2D projection, shadow, or silhouette of the 3D body is obtained. Due to superposition of various tissues, structures, and organs, 2D projection images impose limitations in their interpretation and clinical use. By obtaining a few different projections, such as the commonly used anterior–posterior and lateral projections, the radiologist can overcome some of the limitations of 2D imaging.

In ultrasonography, by using various arrangements of arrays of acoustic transducers, one can obtain sectional views of the body in almost any plane of interest. The image data represent the effects of reflection, transmission, and absorption of the ultrasonic beam by various tissues in the body. The measurements made in ultrasonography may be related to physical characteristics of tissue, such as density and elasticity, and also to motion. Ultrasonography may be used to discriminate between fluid-filled versus solid regions, to analyze the flow of blood based on the Doppler effect, and also to obtain video images of moving parts, such as heart valves. The data obtained for several planes or slices may be combined to generate 3D images of parts or regions of the body. By combining 3D images obtained at several instants of time, it is also possible to generate four-dimensional (4D) datasets as functions of the three spatial coordinates as well as time.

Whereas X-ray and ultrasonographic images facilitate visualization and analysis of various anatomical or physical characteristics of the body, nuclear medicine imaging and MRI facilitate additional depiction of functional or physiological aspects of specific organs or systems. The rate of uptake and discharge of the radionuclide used in modalities of nuclear medicine imaging, such as positron emission tomography (PET) and single-photon emission computed tomography (SPECT), depends upon the rate of flow of blood in the part of the body being examined, and could be used to analyze tumors as well as ischemic, infarcted, or necrotic areas. Various modalities of MRI may be used to obtain anatomical images in terms of different relaxation times of tissues after stimulation by electromagnetic energy. Functional MRI may be used to gather data on physiological characteristics of parts of the body related to the rate of uptake or discharge of oxygen and other material. Time-series of 3D MR images may be combined into 4D datasets.

Medical imaging is important not only for the diagnosis of pathological or abnormal conditions, but also in planning for therapy or surgery. It is common to use simultaneously several images obtained using multiple modalities of medical imaging to investigate and establish the conditions affecting a patient. CT and some modes of MRI can provide images with good resolution of anatomical detail, but may not clearly demonstrate functional, physiological, or pathological aspects. PET, SPECT, and some modes of MRI can demonstrate functional aspects, but with poor spatial resolution. The strengths and advantages of various modalities of medical imaging may be combined to overcome their individual limitations by fusing them together. The integration of anatomical and functional details is of particular interest in diagnosis. For example, functional MR images that have high sensitivity to pathology and specificity may be combined with CT images that offer high spatial resolution and detail to demonstrate anatomical localization and dissemination of malignant disease. Spatial registration of various images is required prior to data fusion. The result of fusion leads to a multivariate or vector-valued image which requires pseudocolor techniques for efficient visualization and analysis [7].

1.3   DATABASES OF MEDICAL IMAGES

Several health regions or jurisdictions require medical images of patients to be archived for periods of about seven years. Images of children may be required to be archived until they are adults. With the practice of obtaining several multimodality images of high resolution and different types of details, a single investigation may generate several gigabytes (GB) of data. Efficient compression of images, preferably with no loss of data or information [3], is essential for cost-effective archival and fast retrieval. Radiology information systems (RIS) and hospital information systems (HIS) face increasing demands on resources for archival and rapid retrieval of medical images, which are being facilitated via PACS and DBMS.

Digital representation of images and patient information calls for methods for indexing and tagging of data files that are substantially different from the conventional style of physical labels made of paper or other material that were affixed to film or other records. New techniques of image processing and pattern recognition facilitate the derivation of numerical or quantitative attributes that represent the visual and diagnostic contents of images. The attributes may be used not only for indexing of images in a database but also for CAD and content-based image retrieval (CBIR): this is one of the major topics of concern in the present book.

1.4   CBIR OF MEDICAL IMAGES

The term CBIR refers to the retrieval of images from a database that are similar to a query image, using measures of information derived from the images themselves, rather than relying on the accompanying text or annotation [8]. To facilitate CBIR, the contents of the images need to be characterized by quantitative features. In a query or search procedure, the features of the query image are compared with the features of each image in the database, and images having high similarity with respect to the query image are retrieved and displayed [8–12]. CBIR of medical images is a useful tool, and could provide radiologists with assistance in the form of a display of relevant past cases with proven pathology, along with the associated clinical, diagnostic, and other information [10–14].

One of the challenging aspects of CBIR is to extract features from the images to represent efficiently their diagnostic and visual information content [8, 10–12]. Generic features of images may not be suitable for the analysis of similarity of medical images of a specific modality of interest. The present work focuses on mammography, using specific features that have been proven to be suitable for mammographic CBIR [11]. In the context of mammography, some works have explored the use of CBIR. Kinoshita et al. [11] and Azevedo-Marques et al. [12] developed a mammographic image retrieval system for use as a diagnostic aid; they proposed the use of visual features obtained from Haralick's measures of texture [15]. Alto et al. [10] proposed the use of texture, gradient (edge-sharpness), and shape measures as indices for quantitative representation of breast masses in mammograms, and studied their application for CBIR. They suggested that features that can give high accuracy in pattern classification could also be used as efficient indices for CBIR. El-Naqa et al. [9] proposed an approach for the retrieval of digital mammograms based on features of clusters of microcalcifications. They explored the use of neural networks and support vector machines (SVMs) in a two-stage hierarchical learning network to predict perceptual similarity from similarity scores collected in human-observer studies. Muramatsu et al. [16] proposed a psychophysical measure based on neural networks for the evaluation of similar images with mammographic masses. For other related previous works, see Tao and Sklansky [17], Ornes et al. [18], El-Naqa et al. [19], and Nakagawa et al. [20, 21].

Similar to CAD systems, CBIR systems use information extracted from images to represent or describe them; however, the main purpose of CBIR is application related to semantic description of the knowledge extracted from the images. Such items of information could be referred to as "visual features" [11, 12]. Initial systems for image retrieval were text-based, using textual descriptors to represent images, such as annotations, keywords, and descriptions of visual characteristics of color, texture, and shape. In the medical context, textual characteristics appear in the form of information related to patient records. In an automatic system, image characteristics could be extracted in the form of measures, features, or attributes [10–12]. In a CBIR system, images are indexed by vectors of characteristic features extracted from the images. In order to retrieve images from a database in response to a query, a comparison is performed between the feature vector of the query image and the corresponding vectors of the images in the database. The comparison is made using measures of similarity in the space of metrics related to the features. Several functions exist for use as measures of similarity of feature vectors. It is important to note that comparison of images is not performed based on their pixel values.

Although CBIR systems have been proposed for several applications, one encounters a "semantic gap" between the quantitative features used to represent the images and the interpretation of the images by users who are experts in the domain of application [8, 12, 22]. This leads to the need to guide the retrieval algorithm by incorporating the user's judgment of similarity and the relevance of each retrieved image with respect to the query image [12, 23, 24]. One of the objectives of current research is to improve the precision of retrieval in a mammographic CBIR system by embedding relevance feedback (RFb) techniques [12, 25] in the retrieval algorithm, and to validate the methods with expert radiologists specialized in mammography.

1.5   REMARKS

The use of multiple modalities of digital imaging in medical diagnostics has created new challenges in their clinical use, interpretation, and management. A CBIR system uses quantitative features for indexing, comparative analysis, and retrieval of images from a database by computing similarity measures with respect to the features of a query image. A CAD system may use similar measures and methods, but with the aim of leading toward a diagnostic decision. Regardless, both CBIR and CAD systems may be used as diagnostic aids. The remaining chapters of the book present methods that serve these purposes.

CHAPTER 2

Mammography and CAD of Breast Cancer

2.1 MAMMOGRAPHY FOR THE DETECTION OF BREAST CANCER

Cancer is caused when a single cell or a group of cells escapes from the controls that regulate cellular growth, and begins to multiply and spread [26]. Such activity results in a mass, tumor, or neoplasm. Some masses are benign: the abnormal growth is restricted to a single, well-circumscribed, expanding mass of cells. Some tumors are malignant: the abnormal growth invades the surrounding tissues and may spread, or metastasize, to other areas of the body [26].

Although curable, especially when detected at early stages, breast cancer is a major cause of death in women. Whereas the cause of breast cancer has not yet been fully understood, early detection and removal of the primary tumor are essential and effective methods to reduce mortality [26].

Mammography is the method of choice for periodic screening of asymptomatic women for early detection of breast cancer [27]. Mammography has gained recognition as the most successful technique for the detection of early, clinically occult breast cancer [27–31]. Screening programs generate large numbers of mammograms that must be viewed and interpreted by the usually limited number of expert radiologists available. The associated heavy workload calls for the application of CAD systems to facilitate efficient analysis of mammograms and accurate diagnosis.

Breast density has been shown to be a risk factor in the development of breast cancer. Wolfe [32] presented the first study relating the density and structure of breast tissues as seen on mammograms to the characteristics of breast disease: he described and illustrated cases associating patterns of parenchymal distortion with the risk of development of breast cancer. Since then, several other researchers have studied the relation between the structural composition of breast tissue and the abnormalities found in the related regions [33]. A consequence of the understanding of this relationship has been the development of systems for the description and analysis of the density patterns found in mammograms: the Breast Imaging–Reporting and Data System (BI-RADS^{TM}), developed by the American College of Radiology (ACR) [34], is the most important of such systems. BI-RADS contains recommendations for standardization of terms used in image-based diagnosis of breast diseases, the division of breast composition and mammographic findings into categories, and suggestions for further actions by the radiologist.

Visual analysis of mammograms takes into consideration the shape and size of the breast, the conditions of the breast contour and the nipple position, and the distribution of fibroglandular tissue (degree of granularity, amount, and distribution of breast density). The characteristics of any dense regions, masses or tumors, calcifications, and distortion in the architecture of the breast parenchyma are also analyzed in detail.

Notwithstanding the developments mentioned above, visual analysis of mammograms by radiologists remains subjective, and suffers from a high degree of intraobserver and interobserver variability [35]. A given mammogram may be categorized into different classes by various radiologists, or even by the same radiologist at different instants of time. With the aim of reducing such variability, systems have been proposed for CBIR and CAD based upon objective measures that represent, in a quantitative manner, mammographic features related to breast structure, diseases, and diagnosis [2, 10–13, 35–38].

2.2   CHARACTERISTICS OF MAMMOGRAMS

2.2.1   X-RAY IMAGING

Soon after the discovery X rays of by Röntgen in 1895, the medical diagnostic potential of the newfound form of radiation was realized. (See Robb [39] for a review of the history of X-ray imaging.) Planar radiography is the simplest form of X-ray imaging, in which a 2D projection, shadow, or silhouette of a 3D body is produced by irradiating the body with an X-ray beam [3–6, 40, 41]. Each ray of photons of the X-ray beam is attenuated to an extent depending upon the integral of the linear attenuation coefficient (μ) along the path of propagation of the ray, and produces a corresponding signal, image value, or gray level at the related point on the detector used.

Considering the ray AB in Figure 2.1, let N_i denote the number of X-ray photons incident upon the body being imaged, within a specified time interval. Assume that the X rays are mutually parallel, with the X-ray source at a large distance from the subject or object being imaged. Let N_o be the corresponding number of photons exiting the body. Then,

$$N_o = N_i \exp\left[-\int_{\mathrm{rayAB}} \mu(x, y, z)\, ds\right], \tag{2.1}$$

or

$$\int_{\mathrm{rayAB}} \mu(x, y, z)\, ds = \ln\left(\frac{N_i}{N_o}\right). \tag{2.2}$$

These equations are modified versions of Beer's law or the Beer-Lambert law on attenuation of X rays. The variable of integration, ds, represents the elemental distance along the ray. The integral is evaluated along the ray AB from the X-ray source to the detector.

The quantities N_i and N_o are Poisson variables; it is assumed that their values are large. The function $\mu(x, y, z)$ represents the linear attenuation coefficient at the point (x, y, z) in the body being imaged. The value of $\mu(x, y, z)$ depends upon the density of the body or object as well as the energy (wavelength or frequency) of the X rays used. In Equation 2.2, it is assumed that monochromatic or monoenergetic X rays are used.

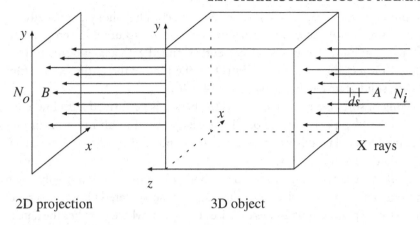

2D projection   3D object

Figure 2.1: Mathematical representation of an X-ray image or a planar radiograph as a 2D projection of a 3D object.

A measurement of the number of photons in the exiting X-ray beam (that is, N_o, and N_i for reference) thus gives us only an integral of $\mu(x, y, z)$ over the corresponding ray. The internal details of the body along the path of the ray are compressed to a single point on the image. Extending the same argument to all rays, it is evident that the image obtained is a 2D planar image of the 3D object, where the internal details are superposed. In the case that the rays are parallel to the z axis, as in Figure 2.1, we have $s = z, ds = dz$, and the image obtained is given by

$$g(x, y) = \int \mu(x, y, z) \, dz. \tag{2.3}$$

It is clear that the 3D object or body being imaged is reduced or integrated to a 2D image by the process of planar radiographic imaging.

Until about the year 2000, the most commonly used detector in X-ray imaging was the screen-film combination [6, 41]. In such a system, the X rays exiting from the body being imaged strike a fluorescent (phosphor) screen made of compounds of rare-earth elements, such as lanthanum oxybromide or gadolinium oxysulfide. The X-ray photons are converted into visible-light photons by the screen. The typical thickness of the phosphor layer in screens is in the range $40 - 100 \ \mu$m. A thicker screen provides a higher efficiency of conversion of X rays to light than a thinner screen, but causes more blurring and results in poorer spatial resolution [3]. A light-sensitive film placed in contact with the screen (in a light-tight cassette) captures the result as an image. The film contains a layer of silver-halide emulsion with a thickness of about $10 \ \mu$m. The exposure or blackening of the film depends upon the number of light photons that reach the film.

A fluoroscopy system uses an image intensifier and a video camera in place of the film to capture the image and to display it on a monitor as a movie or video [6, 41]. Images are acquired at a rate of $2 - 8$ frames/s, with the X-ray beam pulsed at $30 - 100$ ms per frame.

In computed radiography (CR), a photo-stimulable phosphor plate, usually made of europium-activated barium fluorohalide, is used instead of film to capture and temporarily hold the image pattern. The latent image pattern is subsequently scanned with a laser beam and the resulting signal is digitized. In direct digital radiography (DR), the film or the entire screen-film combination is replaced with solid-state electronic detectors [42–46].

Imaging of soft-tissue organs, such as the breast, is performed with low-energy X rays, in the range of $25 - 32$ kVp [47]. The use of a higher kVp would result in lower differential attenuation and poorer tissue-detail visibility or contrast. A focal spot size of $0.1 - 0.3$ mm is desired to maintain high image sharpness in mammography.

As an X-ray beam propagates through a body, photons are lost not only due to absorption but also scattering at each point in the body. The angle of the scattered photon is a random variable, and hence the scattered photon contributes to noise at the point where it strikes the detector. Furthermore, scattering results in the loss of contrast of the part of the object where X-ray photons were scattered from the main beam. The effect of scatter may be reduced by the use of grids.

The interaction between X rays and a detector is governed by the same rules as for interaction with any other matter [5, 40, 41]. Photons are lost due to scatter and absorption; some photons may pass through unaffected or undetected. The small size of the detectors in DR imaging reduces their detection efficiency and causes increased noise. Scattered and undetected photons also cause noise.

2.2.2   MAMMOGRAPHIC IMAGING

A schematic representation of a mammographic imaging system is shown in Figure 2.2. Mammography requires a high-quality, narrow-band (nearly monochromatic) X-ray beam. Molybdenum is the commonly used target material in the X-ray tube or source; molybdenum is also used for beam filtration.

Effective breast compression is an important part of the imaging protocol, designed to eliminate motion, separate mammary structures, create as uniform a density distribution as possible, reduce scattered radiation, and increase the visibility of details in the image. The use of grids specifically designed for mammography can reduce scattered radiation and improve subject contrast, which is especially significant when imaging thick or dense breasts [48].

Generally, mammographic imaging is performed with the breast directly in contact with the detector, resulting in life-size images. In the imaging technique for magnification mammography, an air gap is interposed between the breast and the detector, so that the projected image is enlarged. Magnification produces images containing fine anatomical details that may be useful, especially in cases where conventional mammographic imaging demonstrates uncertain or equivocal findings [49]. The advantages of imaging using grids and magnification are achieved at the cost of increased radiation exposure.

Typically, two mammographic views of each breast are obtained in screening programs: the craniocaudal (CC) view, with the breast compressed and projected along the vertical axis of the body, and the mediolateral-oblique (MLO) view, with the breast compressed and projected from the

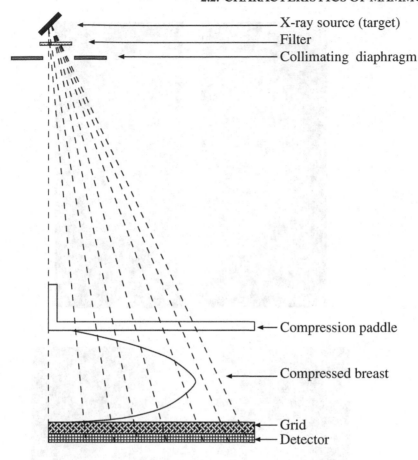

X-ray source (target)

Filter

Collimating diaphragm

Compression paddle

Compressed breast

Grid

Detector

Figure 2.2: A schematic representation of a mammography setup.

middle of the chest toward the side at an oblique angle. Figure 2.3 shows the two standard views of each breast of a normal subject. The structures of the breast form an oriented texture [3] converging toward the nipple.

2.2.3   DIAGNOSTIC FEATURES IN MAMMOGRAMS

Mammograms depict projected images of various structures of the breast, including ducts, vessels, fibroglandular tissue, ligaments, fatty regions, and the skin. The projected nature of mammograms causes the various structures at different distances along the path of propagation of the X rays to be superposed, which, in turn, leads to ambiguity in interpretation. Any disturbance from the normal oriented patterns converging toward the nipple may be cause for suspicion.

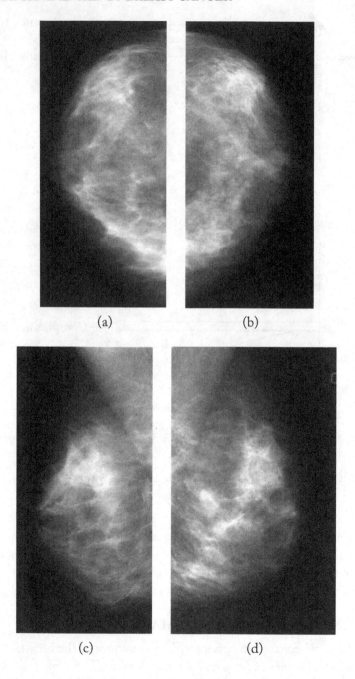

Figure 2.3: CC views of (a) the right breast and (b) the left breast, and MLO views of (c) the right breast and (d) the left breast of a normal subject.

There is substantial evidence of increased occurrence of breast cancer in mammographically dense tissue [50, 51]. Several works have been directed to address the problem of quantification of breast density and its association with the risk of breast cancer [32, 52–59]. Most of such works propose an index or a set of values for the quantification of breast tissue density.

Mammograms are categorized according to breast tissue composition, using the BI-RADS ranking system, as follows [34, 60]:

1. the breast is almost entirely composed of fat, with less than 25% being glandular tissue;

2. 25 − 50% of the breast is composed of scattered fibroglandular tissue;

3. 51 − 75% of the breast is composed of heterogeneously dense breast tissue; and

4. the breast is extremely dense, with more than 75% of the composition being glandular tissue.

Figure 2.4 shows examples of mammograms in the four BI-RADS categories. The histograms of the images are shown in Figure 2.5. The digitized and quantized pixel values in the four images, by necessity, had to be transformed to the range [0, 255] to derive the histograms, which was performed by linear mapping. In order to suppress the large peak that would be caused by the near-zero and low values in the background regions outside the breast portions of the images, the values in the histograms in the range [0, 30] were suppressed in the plots. In general, the histograms show larger counts of pixels in higher density ranges from the BI-RADS 1 to the BI-RADS 4 images. However, the histograms of the BI-RADS 2 and BI-RADS 3 images do not clearly exhibit the expected trends. Close inspection of the original images in Figure 2.4 reveals that the distinction is not clear in the mammograms as well. The examples illustrate the difficulty in manual ranking and also the potential for the scanning process and gray-scale transformations applied to the images to cause ambiguity in ranking. It should be observed that screen-film mammograms are not typically calibrated during acquisition and hence their intensities or densities cannot be compared directly.

Some of the signs of abnormality that may be seen in mammograms include the following [3, 27, 37, 38]:

- deposits of calcium (calcification);

- cysts, masses, and tumors;

- asymmetric arrangement of density or vascular patterns between the left and right breasts; and

- architectural distortion.

Each of the signs listed above could be related to cancer (malignant disease) or other breast diseases that are benign (not malignant). Some of the signs may be difficult to detect due to super-position by other structures of the breast in the projected mammographic image. It is also possible for patterns of superposed normal structures to mimic the appearance of signs of abnormality.

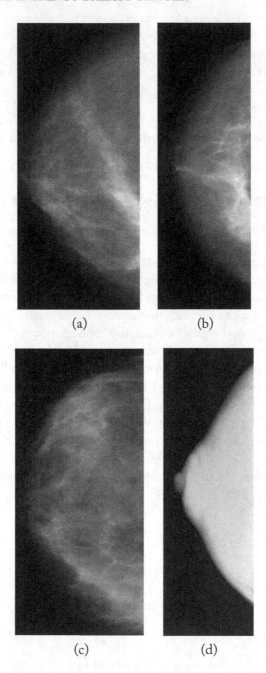

(a)   (b)

(c)   (d)

Figure 2.4: (a)–(d) Mammograms corresponding to BI-RADS density-based categories of 1 to 4, in order.

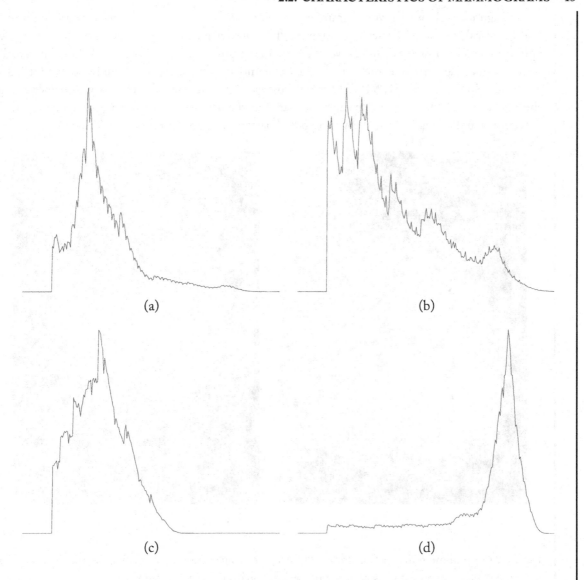

Figure 2.5: (a)–(d) Histograms of the mammograms corresponding to BI-RADS density-based categories of 1 to 4, in order, shown in Figure 2.4. The range of the gray levels (horizontal axis) is [0, 255] in each case; however, histogram values in the range [0, 30] have been suppressed to prevent the effect of the background regions in the images. The axis labels have been removed to prevent clutter.

Figure 2.6 shows a mammogram with benign calcifications and another mammogram with malignant calcifications. Typically, benign calcifications have smooth and round or oval shape and appear in small numbers within a given breast. On the other hand, malignant calcifications usually have rough or jagged shape and appear in large numbers in clusters; they could also exhibit linear and branching patterns [61, 62]. Although most calcifications are relatively easy to detect due to the much higher density of calcium as compared to the density of normal breast structures, their detection could be made difficult by superposed patterns of dense breast tissue [62, 63].

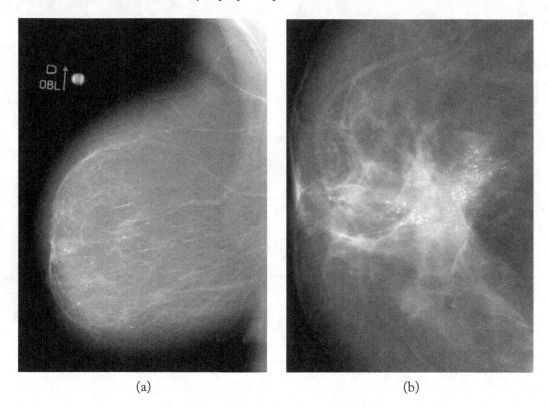

(a)   (b)

Figure 2.6: Mammograms demonstrating (a) benign calcifications and (b) malignant calcifications. The image in (b) is a magnified view of a part of the breast with microcalcifications.

Figure 2.7 shows a mammogram with a benign mass and another mammogram with a malignant tumor. Typically, a benign mass has a smooth and round or oval shape with a well-defined or well-circumscribed border; some benign masses may be macrolobulated [3, 10, 64, 65]. On the other hand, a malignant tumor usually has a rough or jagged shape with an ill-defined, fuzzy, or indistinct border. It is common for malignant tumors to be spiculated, with several strands of high-density tissue patterns appearing to radiate or emanate from the core of the tumor; some malignant tumors may be microlobulated. Notwithstanding the typical characteristics described above, some benign masses could have spicules, and some malignant tumors may have a well-circumscribed border; such cases cause difficulties in the analysis and classification of masses and tumors [3, 10, 64, 65]. Superposed patterns of normal breast tissue could also mimic the appearance of dense regions and spicules, and thereby confound the process of detection and classification of masses or tumors [66].

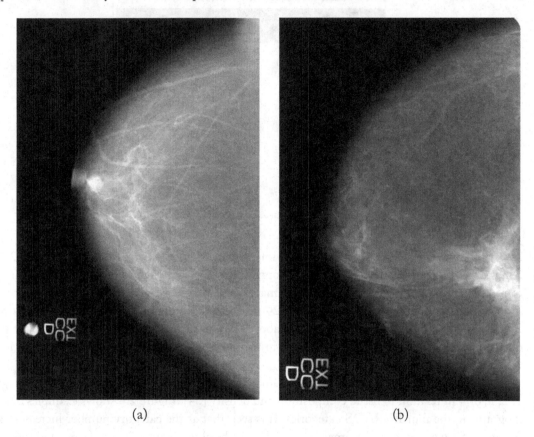

(a)   (b)

Figure 2.7: Mammograms demonstrating (a) a benign mass and (b) a malignant tumor.

2.3   DATABASE OF MAMMOGRAMS

A set of 1080 mammograms was prepared for the present work, including 270 pairs (left and right breasts of a patient) of CC views and 270 pairs of MLO views from the Medical Center of the Faculty of Medicine, University of São Paulo, Ribeirão Preto, São Paulo, Brazil. Approval was obtained from the Research Ethics Committee. The details of the images in the database are summarized in Table 2.1.

The films were digitized using a Vidar DiagnosticPro scanner with a resolution of 300 dpi (pixel size of 85 μm) and 12 bits per pixel (bpp). Films of two sizes are present in the database:

Table 2.1: Numbers of images in the database used in the present work, categorized by radiologists using BI-RADS^{TM} categories 1, 2, 3, or 4 based on tissue composition.

View	1	2	3	4	Total
CC right	90	112	50	18	270
CC left	91	111	50	18	270
MLO right	90	112	50	18	270
MLO left	91	111	50	18	270
Total	362	446	200	72	1080

24×30 cm and 18×24 cm. The digitized image matrices were of size 2835×3543 and 2126×2835 pixels, respectively. The images were cropped to remove patient and imaging markers. In order to reduce computational requirements, the images were down-sampled, without significant loss of information, to matrices of size 1024×1024 pixels, with an effective resolution of 279 μm and 223 μm for the two film sizes mentioned above. The pixel sizes were taken into account in the determination of measures of error, in mm, in the positions of the nipple and the pectoral muscle edge detected.

Mammograms were categorized by radiologists of the hospital using BI-RADS categories 1, 2, 3, or 4 in relation to tissue composition and density [34]. Figure 2.4 shows examples of mammograms in the four BI-RADS categories. It is seen that as the category number increases, larger portions of the mammographic images are covered by brighter regions corresponding to proportionally larger amounts of denser tissues in the breast. This classification based on tissue composition is used in the present work as the "gold standard" for judgment of relevant images for a given query in evaluation of the results of CBIR; see Table 2.1. The BI-RADS classification information is routinely saved in the RIS report, and is referred to as RADR in the related discussions. Images of CC and MLO views were manually separated for the analysis of retrieval.

2.4   CAD OF BREAST CANCER

Several systems for CAD of breast cancer have been designed with the aim of assisting radiologists in the analysis of mammograms [2, 3, 37, 38]. The purpose of CAD is to increase the accuracy of diagnosis as well as to improve the consistency of interpretation of images via the use of the results of processing mammograms with computational procedures as a reference or second opinion [2, 3]. The results of computer-based image processing and CAD could also be useful in addressing other limitations in visual interpretation of mammograms, due to poor quality and low contrast of the images, superposition of breast structures due to the projected nature of mammograms and the compression of the breast for imaging, visual fatigue in the screening context, and environmental distraction. It has been shown that double reading (interpretation of each mammographic exam by two radiologists) can increase the accuracy of diagnosis [2]: the suggested use of CAD systems also includes the role of CAD as a second reader instead of the second radiologist.

CAD systems have well-defined objectives, such as the detection of suspicious lesions, characterization of lesions as benign or malignant, analysis of bilateral asymmetry in breast parenchymal patterns, detection of architectural distortion, and analysis of breast density [3]. An example of the application of a CAD system for the classification of breast density is the work of Zhou et al. [35], which includes an automated system to obtain measures of breast density in accordance with BI-RADS. Classification was realized based upon features extracted from the gray-level histogram and peaks corresponding to fatty and fibroglandular tissues. Nevertheless, it is difficult to achieve high accuracy in deriving measures of breast tissue density due to intrinsic difficulties with mammographic images; furthermore, the estimates provided by radiologists based upon visual analysis are subjective.

An approach to address the problems mentioned above is to realize unsupervised and automatic classification of images through the characterization of similarity based upon breast tissue density. Such a classification permits quantitative evaluation of the similarity of images independent of subjective factors. Classification of images as above may be realized with a CBIR system, the results of which could be evaluated and improved upon via a system of RFb (by expert radiologists specialized in mammography) [12, 36]. Li et al. [13] present an example of applying CBIR as a tool to assist in radiological diagnosis. In this manner, a CAD system can contribute effectively toward the reduction of interobserver variability in the classification of images.

2.5   REMARKS

This chapter presented an overview of mammography as a tool for the detection of breast cancer. The basic radiographic and radiological aspects of mammographic imaging were reviewed. The aspects of mammograms that are important in diagnosis, CAD, and CBIR were described. The notions of CAD of breast cancer and CBIR were presented briefly. Specific details of some of the processes mentioned in this chapter are presented in the remaining chapters of the book.

CHAPTER 3

Segmentation and Landmarking of Mammograms

3.1 OBJECTIVES OF MAMMOGRAPHIC IMAGE PROCESSING

Figure 3.1 shows a typical sequence of the various objectives and steps in computer processing and analysis of mammograms for CAD. Regardless of the level of sophistication or quality of a medical imaging system, it is common to encounter noise and artifacts in the images obtained in a practical setting. Therefore, the first step in a typical CAD system is to remove noise and artifacts. In the present chapter, two methods are described for this purpose: the Wiener filter and anisotropic diffusion.

The second step is to detect and segment ROIs in the given images; some of the ROIs could serve as landmarks that facilitate further processing and analysis of the images. The most useful landmarks in a mammogram are the breast boundary, the nipple, and the chest wall or the boundary of the pectoral muscle, the last one especially in the case of MLO views. Further detailed segmentation steps may be used to outline the fibroglandular disk; dense regions corresponding to masses or tumors; calcifications and their clusters; ducts and vessels; lymph nodes; and subregions corresponding to different tissue types and density levels [3]. Several methods for these purposes are described in the present chapter.

After ROIs are extracted, they need to be characterized objectively in terms of measures of various properties. This step, known as feature extraction, results in quantitative representation of each ROI with a set or vector of features.

In an initial explorative study, one may wish to derive a large number of features to represent various properties of ROIs in many different ways. For the sake of efficiency in the next step of pattern classification, however, it would be desirable to have a compact representation using only the most useful or discriminative subset of features. For this purpose, a step of feature selection is desirable. The result of pattern recognition or classification could then be used to achieve CBIR and CAD.

The following sections and chapters provide details related to the various steps and tasks mentioned above.

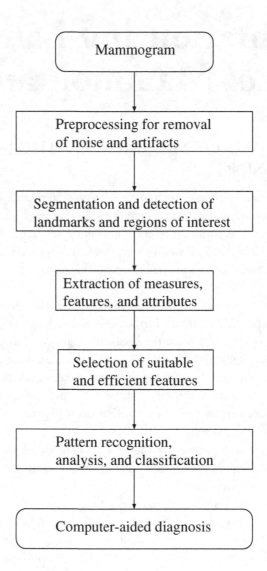

Figure 3.1: Schematic representation of the objectives of computer processing and analysis of mammograms for the purpose of CAD.

3.2   ENHANCEMENT OF MAMMOGRAMS

3.2.1   THE WIENER FILTER

Lee [67] proposed adaptive filters based on local statistics to obtain an estimate of the original image from a degraded version. The model used represents an image, $f(m, n)$, corrupted by additive noise, $\eta(m, n)$, as

$$g(m, n) = f(m, n) + \eta(m, n), \quad \forall\, m, n. \tag{3.1}$$

The original image f is considered to be a realization of a nonstationary random field, characterized by spatially varying moments (such as mean and standard deviation). The noise process η may be either signal-independent or signal-dependent, and could be nonstationary.

In the approach proposed by Lee, an estimate, $\tilde{f}(m, n)$, of the original image, $f(m, n)$, is computed at every spatial location (m, n) by applying a linear operator to the corrupted image $g(m, n)$. Scalars $a(m, n)$ and $b(m, n)$ are sought such that $\tilde{f}(m, n)$, computed as

$$\tilde{f}(m, n) = a(m, n)\, g(m, n) + b(m, n) \tag{3.2}$$

minimizes the local mean-squared error (MSE), defined as

$$\varepsilon^2(m, n) = \overline{\left[\tilde{f}(m, n) - f(m, n)\right]^2}, \tag{3.3}$$

where the bar above the expression indicates averaging (statistical expectation or spatial averaging). The values $a(m, n)$ and $b(m, n)$ that minimize $\varepsilon^2(m, n)$ are computed by taking the partial derivatives of $\varepsilon^2(m, n)$ with respect to $a(m, n)$ and $b(m, n)$, setting them to zero, and solving the resulting equation [3]. The result is

$$\tilde{f}(m, n) = \overline{f}(m, n) + \frac{\sigma_{fg}(m, n)}{\sigma_g^2(m, n)}\, [g(m, n) - \overline{g}(m, n)], \tag{3.4}$$

where σ_{fg} is the covariance between f and g, and σ_g^2 is the variance of g.

Because the true statistics of the original and the corrupted images as well as their joint statistics are unknown in a practical situation, Lee proposed to estimate them locally in a spatial neighborhood of the pixel (m, n) being processed, leading to the local linear minimum MSE (that is, the LLMMSE) estimate. Using a rectangular window of size $(2P + 1) \times (2Q + 1)$ centered at the pixel (m, n) being processed, local estimates of the mean, μ_g, and variance, σ_g^2, of the noisy image g are obtained as

$$\mu_g(m, n) = \frac{1}{(2P + 1)(2Q + 1)} \sum_{p=-P}^{P} \sum_{q=-Q}^{Q} g(m + p, n + q), \tag{3.5}$$

and

$$\sigma_g^2(m, n) = \frac{1}{(2P + 1)(2Q + 1)} \sum_{p=-P}^{P} \sum_{q=-Q}^{Q} [g(m + p, n + q) - \mu_g(m, n)]^2. \tag{3.6}$$

The LLMMSE estimate is then approximated by the following pixel-by-pixel operation:

$$\tilde{f}(m, n) = \mu_g(m, n) + \left[\frac{\sigma_g^2(m, n) - \sigma_\eta^2(m, n)}{\sigma_g^2(m, n)} \right] [g(m, n) - \mu_g(m, n)]. \tag{3.7}$$

Comparing Equation 3.4 with 3.7, it is seen that $\overline{f}(m, n)$ is approximated by $\mu_g(m, n)$, and $\sigma_{fg}(m, n)$ is estimated by the difference between the local variance of the degraded image, σ_g^2, and that of the noise process, σ_η^2. The LLMMSE filter, also known as the Wiener filter, is spatially adaptive and nonlinear due to space-variant estimation of the statistical parameters used.

3.2.2 ANISOTROPIC DIFFUSION

The approach of anisotropic diffusion for image processing, proposed by Perona and Malik [68], facilitates selective smoothing while maintaining contrast and sharpness across the edges present in the image. For this purpose, a locally adaptive gradient measure is used. The model of diffusion used is

$$\frac{\partial f}{\partial t} = \text{div}\{c \, \nabla f\}, \tag{3.8}$$

where t indicates the time variable for the diffusion process, div is the divergence operator, and ∇f is the gradient of the image f. The coefficient of divergence, c, is defined as

$$c = \exp\left[-\left\{ \frac{\|\nabla f\|}{k} \right\}^2 \right], \tag{3.9}$$

where k is a threshold or weight on the gradient. By defining the coefficient of divergence as a function of the local gradient of the image, the diffusion process is made anisotropic and adaptive. When the local gradient is large, such as near edges, the diffusion is kept low so as to avoid blurring across edges. When the local gradient is small, smoothing is allowed to a larger extent.

In discrete implementation, the filtered image is obtained in an iterative procedure using the following equation:

$$f_{t+1} = f_t + \lambda\{c_N \nabla_N f + c_S \nabla_S f + c_E \nabla_E f + c_W \nabla_W f\}, \tag{3.10}$$

where t is the iteration number, and the subscripts N, S, E, and W indicate the North, South, East, and West directions. The gradients in the N, S, E, and W directions are estimated using first derivatives as $f(m-1, n) - f(m, n)$, $f(m+1, n) - f(m, n)$, $f(m, n+1) - f(m, n)$, and $f(m, n-1) - f(m, n)$, respectively. The coefficients c_N, c_S, c_E, and c_W are computed at each iteration using the corresponding gradients. Perona and Malik [68] recommended $0 \leq \lambda \leq 0.25$ for stability.

Catté et al. [69, 70] proposed a modification to the definition of c as

$$c = \exp\left[-\left\{ \frac{\|\nabla(G * f)\|}{k} \right\}^2 \right], \tag{3.11}$$

where $G * f$ represents prefiltering of the image f with a Gaussian lowpass filter function, G, to reduce the effects of noise on the gradient. The variance of G controls the scale of filtering.

Segall and Acton [71] further modified the coefficient of divergence with nonlinear morphological filters as

$$c = \exp\left[-\left\{\frac{\|\nabla((f \bullet S) \circ S)\|}{k}\right\}^2\right], \qquad (3.12)$$

where $\bullet$ and $\circ$ represent morphological closing and opening, respectively, with the structuring element S. The morphological filter limits displacement of edges and smoothens objects smaller than the scale indicated by the structuring element used.

Figure 3.2 shows examples of filtering a mammogram with the anisotropic diffusion filter including prefiltering with a Gaussian filter having unit standard deviation and width of 11 pixels, morphological filtering with the radius of the structuring element equal to 5 pixels, and the Wiener filter with the window size of 11 pixels. The results of filtering were evaluated by computing the signal-to-noise ratio (SNR) as

$$\text{SNR} = 10 \ \log_{10}\left[\frac{\sum_{m=0}^{M-1}\sum_{n=0}^{N-1}\{\tilde{f}(m,n)\}^2}{\sum_{m=0}^{M-1}\sum_{n=0}^{N-1}\{\tilde{f}(m,n) - g(m,n)\}^2}\right], \qquad (3.13)$$

where $\tilde{f}(m,n)$ is the result of filtering $g(m,n)$ and both images are of size $M \times N$ pixels. The results of the three filtering procedures mentioned above, in order, had SNR values of 19.65, 19.03, and 22.26 dB, indicating that the best results were obtained with the Wiener filter followed by anisotropic diffusion.

3.3 SEGMENTATION OF MAMMOGRAMS

CAD systems and methods are used to identify suspicious lesions (calcifications, masses, and distortion of parenchymal tissue) as well as to quantify the density of the breast tissue. To accomplish these goals, an ROI may be extracted from the given mammogram or the entire breast region may be analyzed [35, 52, 72–74]. In some methods, bilateral comparison of mammograms is performed to detect asymmetry [75, 76]. In all of these cases, it is necessary to segment the breast area in the mammogram to identify ROIs [3, 77–79].

When MLO mammograms are used, an additional step to segment the pectoral muscle is desirable because this region appears at approximately the same density as the dense tissues of interest in the image. Therefore, the breast tissue must be isolated from the pectoral muscle and the background [77]. The approximately straight line that separates the region of the pectoral muscle from the breast portion in an MLO view is important also for automatic alignment of mammograms for bilateral analysis and detection of asymmetry; the difference between the mammograms of the left and right breasts of a woman could also be used to detect masses and architectural distortion [74, 80].

Karssemeijer [52] used the Hough transform and a set of threshold values applied to the accumulator cells in the Hough space to detect a straight-line approximation of the edge of the

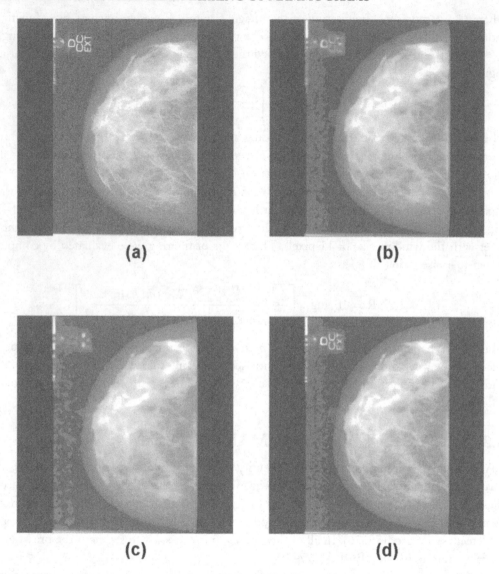

(a)

(b)

(c)

(d)

Figure 3.2: (a) Original mammogram and results of filtering with the anisotropic diffusion filter including prefiltering with (b) a Gaussian filter with standard deviation = 11 pixels, (c) morphological filtering with the radius of the structuring element equal to 5 pixels, and (d) the Wiener filter with the window size of 11 pixels. Reproduced with kind permission from Springer Science+Business Media B. V. from S. K. Kinoshita, P. M. de Azevedo-Marques, R. R. Pereira Jr., J. A. H. Rodrigues, and R. M. Rangayyan, "Radon-domain detection of the nipple and the pectoral muscle in mammograms," Journal of Digital Imaging, 21(1): 37–49, March 2008. © Springer.

pectoral muscle. The Radon transform has been used to locate linear structures in an image [81–83]. Section 3.3.4 presents methods based on the Radon transform for automatic detection of the pectoral muscle region.

Furthermore, the position of the nipple would be useful as an alignment point or landmark for registration of mammograms [80, 83, 84]. Glandular structures of the breast (such as lobules, ducts, and lobes) converge toward the nipple. Radiologists refer to the nipple to compare corresponding regions of the right and left breasts to detect relative anomalies [84]. The characteristics mentioned above are used in the method described in Section 3.3.5 to detect the nipple; the method is implemented in the Radon domain to locate linear structures in an image [81, 82].

3.3.1   SEGMENTATION OF THE BREAST REGION

A mammographic image contains not only an image of the breast but also regions corresponding to the surrounding air and artifacts present around the breast, such as labels identifying the patient and the imaging protocol. In order to limit the data (pixels) used in the analysis of mammograms to the relevant breast portion only, it is necessary to perform segmentation. This step is referred to variously as detection of the skin–air boundary, detection of the breast boundary, or segmentation of the breast region.

In a typical mammogram, the regions with the highest density or brightness correspond to the labels placed on the imaging platform to identify the patient and indicate the imaging protocol (such as CC, MLO, and left or right breast). These labels are expected not to be in contact with the breast. The region with the next highest brightness would typically be the pectoral muscle, especially in MLO views. Calcifications and masses, if present in the breast, could be of brightness comparable to that of the pectoral muscle. The fibroglandular disk contains a combination of breast tissues at various levels of density and low-density fat. The histogram of the breast portion of a mammogram will be skewed to the left (dark values) if a large portion of the breast is composed of fatty and low-density tissues. As the relative amount of higher density tissue increases, the histogram gets skewed more to the right (bright values); see Figure 2.5. The histogram is useful in guiding the segmentation procedure to identify the breast region.

In the present work, after filtering to remove noise as described in Section 3.2, segmentation of the breast region was performed as follows [83]. Several thresholding methods were applied to each image, including the maximum-entropy principle [85], a moment-preserving method [86], Otsu's method [87], the method of Ridler and Calvard [88], the method of Reddi et al. [89], and a method based upon the gray-level cooccurrence matrix (GLCM) [90]. The best result among the above was selected for each image by one of the radiologists involved in the study. Three of the methods mentioned above are described in the following paragraphs.

In Otsu's method [87], the probability density function (PDF) of gray levels, $p(l), l = 0, 1, 2, \ldots, L - 1$, is divided into two classes C_0 and C_1 separated by a threshold k. The probabilities of occurrence ω_i, $i = \{0, 1\}$, of the classes C_i, $i = \{0, 1\}$, are given by

$$\omega_0(k) = P(C_0) = \sum_{l=0}^{k} p(l) = \omega(k) \tag{3.14}$$

and

$$\omega_1(k) = P(C_1) = \sum_{l=k+1}^{L-1} p(l) = 1 - \omega(k). \tag{3.15}$$

The class mean levels μ_i for C_i, $i = \{0, 1\}$, are given by

$$\mu_0(k) = \sum_{l=0}^{k} l \, P(l|C_0) = \sum_{l=0}^{k} l \, \frac{p(l)}{\omega_0(k)} = \frac{\mu(k)}{\omega(k)}, \tag{3.16}$$

and

$$\mu_1(k) = \sum_{l=k+1}^{L-1} l \, P(l|C_1) = \sum_{l=k+1}^{L-1} l \, \frac{p(l)}{\omega_1(k)} = \frac{\mu_T - \mu(k)}{1 - \omega(k)}, \tag{3.17}$$

where

$$\omega(k) = \sum_{l=0}^{k} p(l) \tag{3.18}$$

and

$$\mu(k) = \sum_{l=0}^{k} l \, p(l) \tag{3.19}$$

are the cumulative probability and first-order moment of the PDF $p(l)$ up to the threshold level k, and

$$\mu_T = \sum_{l=0}^{L-1} l \, p(l) \tag{3.20}$$

is the mean gray level of the image.

The class variances are given by

$$\sigma_0^2(k) = \sum_{l=0}^{k} [l - \mu_0(k)]^2 \, P(l|C_0) = \sum_{l=0}^{k} [l - \mu_0(k)]^2 \, \frac{p(l)}{\omega_0(k)}, \tag{3.21}$$

and

$$\sigma_1^2(k) = \sum_{l=k+1}^{L-1} [l - \mu_1(k)]^2 \, P(l|C_1) = \sum_{l=k+1}^{L-1} [l - \mu_1(k)]^2 \, \frac{p(l)}{\omega_1(k)}. \tag{3.22}$$

Using the discriminant criterion

$$v = \frac{\sigma_B^2(k)}{\sigma_T^2},$$ (3.23)

where

$$\sigma_B^2(k) = \omega_0(k)[\mu_0(k) - \mu_T]^2 + \omega_1(k)[\mu_1(k) - \mu_T]^2$$ (3.24)

and

$$\sigma_T^2 = \sum_{l=0}^{L-1} (l - \mu_T)^2 p(l),$$ (3.25)

Otsu's algorithm finds the threshold level k that maximizes the discriminant criterion v given in Equation 3.23. Maximizing v is equivalent to maximizing σ_B^2, because the value σ_T^2 does not vary with the threshold value k. The optimal threshold value k^* is given as

$$k^* = \arg \left\{ \max_{0 \leq k \leq L-1} \sigma_B^2(k) \right\}.$$ (3.26)

In the method of Kapur et al. [85], two measures of entropy as computed as follows:

$$H_b(k) = -\sum_{l=0}^{k} \frac{p(l)}{P_k} \log_2 \left[\frac{p(l)}{P_k} \right]$$ (3.27)

and

$$H_w(k) = -\sum_{l=k+1}^{L-1} \frac{p(l)}{1-P_k} \log_2 \left[\frac{p(l)}{1-P_k} \right],$$ (3.28)

where

$$P_k = -\sum_{l=0}^{k} p(l).$$ (3.29)

The optimal threshold is the value of k that maximizes $H_b(k) + H_w(k)$.

The procedure of Ridler and Calvard [88] is iterative and works as follows. An initial index, k, is set equal to the average gray level in the given image. The average value of all of the pixels in the image lower than k, labeled as μ_1, and the average value of all of the pixels in the image higher than k, labeled as μ_2, are calculated. A threshold value is calculated as $T = (\mu_1 + \mu_2)/2$. If $k < T$, the index, k, is updated to the next higher gray level present in the image, and a new threshold is calculated as above. The procedure is iterated until a stable value of the threshold is obtained.

Figure 3.3 shows a mammogram and the result of segmentation using each of the thresholding methods mentioned above. For several of the images processed, many of the methods mentioned above provided the same or similar threshold values. The results indicate that the characteristics of the images agree with the models and assumptions used in the methods. In further postprocessing steps, morphological closing and opening operations [91, 92] were applied, with circular structuring elements of radius equal to 3, 5, and 11 pixels, to remove small artifacts in the image and to smoothen the contour of the breast region. Subsequently, B-spline interpolation was applied using 32-pixel control-point intervals to obtain the final contour of the breast region.

To obtain the effective breast region from a given mammogram, the result of segmentation as above was used in the form of a mask and multiplied with the original mammogram. Figure 3.4 shows the results of the various steps in the segmentation of a mammogram.

As reported by Kinoshita et al. [83], the results of segmentation of the breast region were evaluated by a radiologist. The acceptance grade was determined in relation to the amount of missing or increased breast tissue apparent in the image. The results indicated that, for CC images, the results were not acceptable for seven (1.3%) images, acceptable for 156 images (28.9%), and accurate for 377 images (69.8%). For MLO views, the results were not acceptable for 21 images (3.9%), acceptable for 154 images (28.5%), and accurate for 365 images (67.6%). In this evaluation, the term "accurate" meant that the segmentation was almost perfect; "acceptable" meant that some breast tissue was excluded or some nonbreast part was included in the segmented result, but the final image was acceptable for diagnosis; and "not acceptable" meant that the final image was not acceptable for diagnosis.

3.3.2 SEGMENTATION OF THE FIBROGLANDULAR DISK

The results of segmentation of the breast region were subjected to further segmentation to obtain the fibroglandular disk [11, 79, 93]. The methods based on the maximum-entropy principle [85], the moment-preserving method [86], and Otsu's method [87], described in Section 3.3.1, were explored for this purpose. Figure 3.5 shows two examples of segmentation of the fibroglandular disk, including a CC view and an MLO view. The results of segmentation include regions with fibroglandular tissues of different levels of density but not regions containing only fat.

3.3.3 THE RADON TRANSFORM

The Radon transform of an image is the projection (integral) of the image at a given angle [3, 94]. Each point in the Radon domain, (t, θ), corresponds to the integration of the image along a straight line (ray) in the spatial domain (x, y) of image, defined as

$$R(t, \theta) = \int_{-\infty}^{\infty} \int_{-\infty}^{\infty} f(x, y) \, \delta(x \cos \theta + y \sin \theta - t) \, dx \, dy. \tag{3.30}$$

The Radon space or domain may be represented by the variables $t = x \cos \theta + y \sin \theta$, which is the position of the ray of integration, and $s = -x \sin \theta + y \cos \theta$, which is the variable indicating distance along the ray. The Radon domain may also be represented by the variables (t, θ); then, $R(t, \theta)$ is expressed as a function of the angle of a ray, θ, and its position or displacement, t. Figure 3.6 shows the relationship between the variables (x, y) of the image domain and the variables (s, t, θ) in the Radon domain.

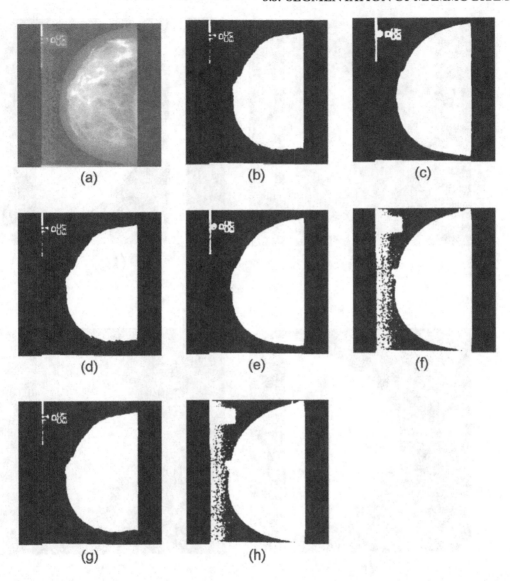

Figure 3.3: Examples of the results of the segmentation of the effective breast region with various methods: (a) original image; (b) maximum-entropy principle; (c) moment-preserving method; (d) Otsu's method; (e) the method of Ridler and Calvard; (f) the method based on the GLCM; (g) and (h) maximum and minimum threshold of the method of Reddi et al. Reproduced with kind permission from Springer Science+Business Media B. V. from S. K. Kinoshita, P. M. de Azevedo-Marques, R. R. Pereira Jr., J. A. H. Rodrigues, and R. M. Rangayyan, "Radon-domain detection of the nipple and the pectoral muscle in mammograms," Journal of Digital Imaging, 21(1): 37–49, March 2008. © Springer.

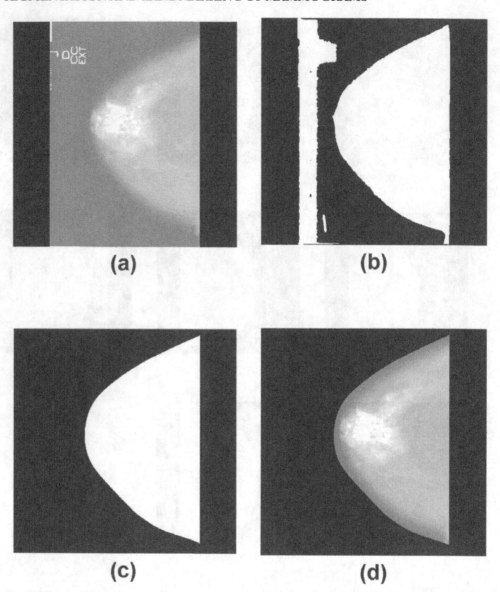

Figure 3.4: (a) Original mammogram and the results of segmentation of the breast region at various stages: (b) initial segmentation, (c) after removing spurious regions and artifacts as well as smoothening the edges, and (d) final result of segmentation of the mammogram. Reproduced with kind permission from Springer Science+Business Media B. V. from S. K. Kinoshita, P. M. de Azevedo-Marques, R. R. Pereira Jr., J. A. H. Rodrigues, and R. M. Rangayyan, "Radon-domain detection of the nipple and the pectoral muscle in mammograms," Journal of Digital Imaging, 21(1): 37–49, March 2008. © Springer.

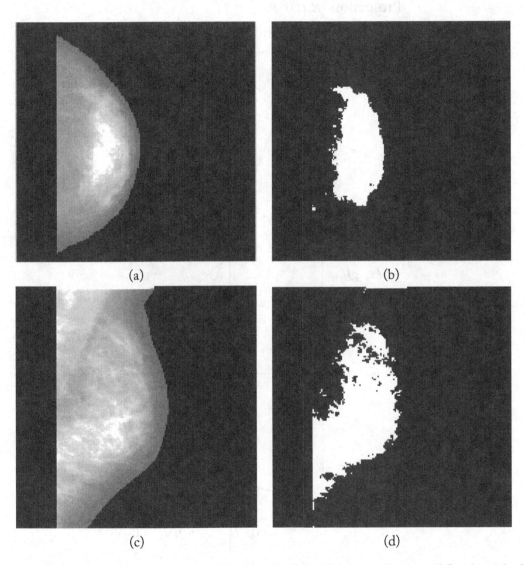

(a)　　　　　　　　　　　　　　　　(b)

(c)　　　　　　　　　　　　　　　　(d)

Figure 3.5: Examples of segmentation of the fibroglandular disk by application of Otsu's method of thresholding [93]: (a) original image, CC view; (b) segmented fibroglandular disk; (c) original image, MLO view; (d) segmented fibroglandular disk. Figure courtesy of S. K. Kinoshita [93].

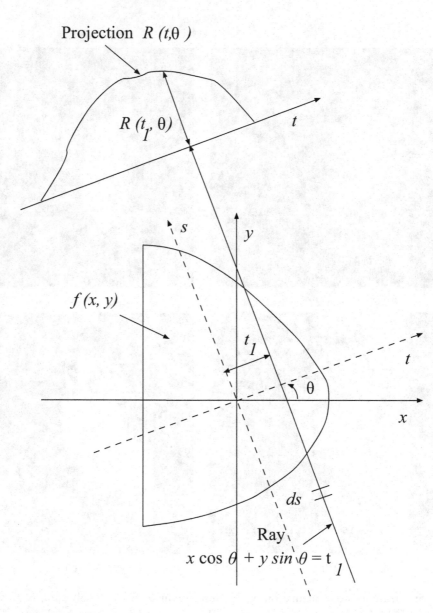

Figure 3.6: Relationship between the variables (x, y) of the image domain and the variables (s, t, θ) in the Radon domain.

The Radon transform forms the basis of the representation of images by their projections (ray integrals), as well as image reconstruction from projections and computed tomography [3, 39, 94]. Several algorithms for image processing and analysis may be applied advantageously in the Radon domain, for applications such as the detection of edges [82], analysis of oriented patterns [81], analysis of spicules and architectural distortion in mammograms [95], and deconvolution via the complex cepstrum [96, 97].

In the present work, the Radon transform is applied to two types of images: binary images for the detection of straight lines (to detect the edge of the pectoral muscle, as described in Section 3.3.4) and gray-level images for the detection of maximum-intensity peaks for selected projections at specific ray positions and directions (to detect the nipple, as described in Section 3.3.5).

Figure 3.7 shows the gray-scale and binary versions of a test image and their Radon transforms represented as 2D arrays or images. Each row in the representation of a Radon transform as an image represents one projection function, $R(t, \theta)$, for a particular value of $t = t_1$. Figure 3.8 shows a CC mammogram as well as the corresponding fibroglandular disk segmented into its external and internal subregions; the Radon transforms of the subregions and functions derived thereof are also shown as images. In addition, the functions $R(t)$ and $R(\theta)$ obtained by integrating $R(t, \theta)$ over θ and t, respectively, are shown. As evident from the examples in Figures 3.7 and 3.8, the Radon transform displays large values at ray angles and positions that correspond to strong linear features in the image.

3.3.4 DETECTION OF THE PECTORAL MUSCLE

The pectoral muscle region has a predominantly high density that could interfere with the analysis of mammograms: prior detection and removal of this region could lead to improved results with CAD algorithms [77]. The high density of the pectoral muscle results in a strong edge that can be approximated by a straight line. Kinoshita et al. [83] proposed a method based on the Radon transform [3, 81, 82] to detect the edge of the pectoral muscle region, which is described in the following paragraphs.

The Canny filter [98] based on a Gaussian profile and gradient estimation was applied for edge detection. The standard deviation of the Gaussian function was varied from 1 to 10 pixels. The Radon transform of the edge image was computed. To identify the pectoral muscle region, a straight line was used to approximate the separation between the breast and pectoral muscle tissues. Straight-line candidates were detected by applying the Radon transform in the angle interval between 5° and 50° for images of the right breast, and between −5° and −50° for images of the left breast. The criteria of maximum intensity in the Radon domain and localization close to the thoracic wall were applied to detect a straight line to represent the edge of the pectoral muscle region. Figure 3.9 presents an example illustrating various stages of the procedure for the detection of the straight-line approximation to the pectoral muscle edge in a mammogram.

Kinoshita et al. [83] tested the method for the detection of the pectoral muscle edge with 540 MLO mammographic images. To evaluate the result of automatic detection of the pectoral muscle edge, a radiologist specialized in mammography independently determined the pectoral muscle edge

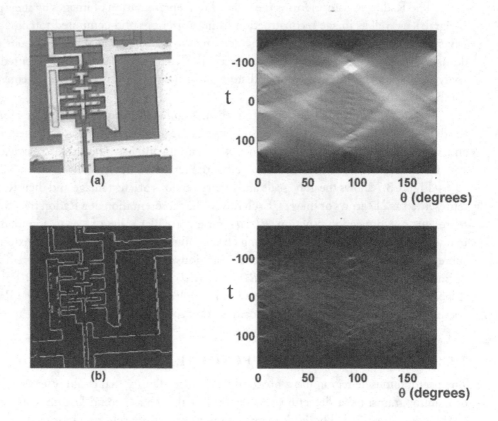

Figure 3.7: Examples of (a) a gray-level image (left) and (b) its binarized version (left) along with their Radon transforms (right). Reproduced with kind permission from Springer Science+Business Media B. V. from S. K. Kinoshita, P. M. de Azevedo-Marques, R. R. Pereira Jr., J. A. H. Rodrigues, and R. M. Rangayyan, "Radon-domain detection of the nipple and the pectoral muscle in mammograms," Journal of Digital Imaging, 21(1): 37–49, March 2008. © Springer.

(approximated by a straight line) by visual analysis. The automatically detected edges were evaluated by using the percentages of false-positive (FP) and false-negative (FN) pixels normalized with reference to the corresponding numbers of pixels in the pectoral muscle region manually delimited by the radiologist. Of the 540 images processed, the results were considered to be accurate (FP and FN < 5%) for 156 images (28.9%); acceptable (FP and FN > 5% but ≤ 15%) for 220 images (40.7%); and not acceptable (FP and FN > 15%) for 164 images (30.4%). The mean and standard deviation values of the FP and FN rates as well as the Hausdorff distance between the automatically detected and manually delimited pectoral muscle edges were computed. The mean and standard

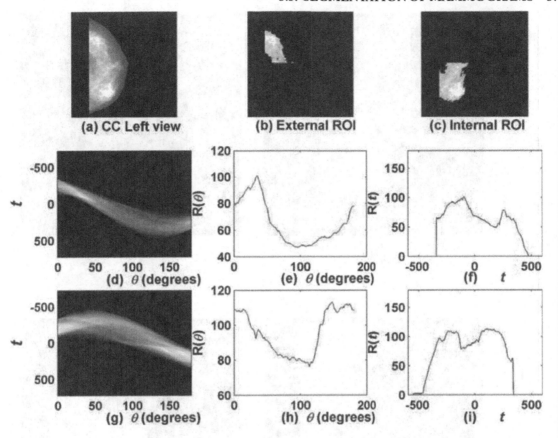

Figure 3.8: Examples of regions of a mammogram and their Radon transforms. (a) An original mammogram, (b) its external ROI, and (c) its internal ROI. Radon transform functions (d) $R(t, \theta)$, (e) $R(\theta)$, and (f) $R(t)$ for the image in (b). Radon transform functions (g) $R(t, \theta)$, (h) $R(\theta)$, and (i) $R(t)$ for the image in (c). Reproduced with kind permission from Springer Science+Business Media B. V. from S. K. Kinoshita, P. M. de Azevedo-Marques, R. R. Pereira Jr., J. A. H. Rodrigues, and R. M. Rangayyan, "Content-based retrieval of mammograms using visual features related to breast density patterns," Journal of Digital Imaging, 20(2): 172–190, June 2007. © Springer.

deviation values of the FP and FN rates for the 540 images tested were 8.99 ± 38.72% and 9.13 ± 11.87%, respectively. The mean and standard deviation values of the Hausdorff distance were 12.45 ± 22.96 mm. Figures 3.10 and 3.11 illustrate two examples of pectoral muscle edge detection considered to be accurate (FP = 1.2%, FN = 3.0%, FP and FN combined = 4.2%) and acceptable (FP = 4.2%, FN = 9.9%, FP and FN combined = 14.1%), respectively.

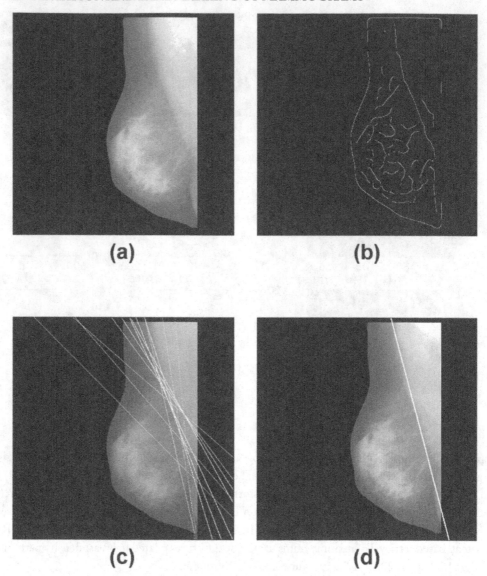

Figure 3.9: Example of the detection of the edge of the pectoral muscle, approximated by a straight line: (a) original image; (b) the edge image obtained by using the Canny filter; (c) detection of straight-line candidates by the Radon transform; and (d) selection of the pectoral muscle edge by the criteria adopted. Reproduced with kind permission from Springer Science+Business Media B. V. from S. K. Kinoshita, P. M. de Azevedo-Marques, R. R. Pereira Jr., J. A. H. Rodrigues, and R. M. Rangayyan, "Radon-domain detection of the nipple and the pectoral muscle in mammograms," Journal of Digital Imaging, 21(1): 37–49, March 2008. © Springer.

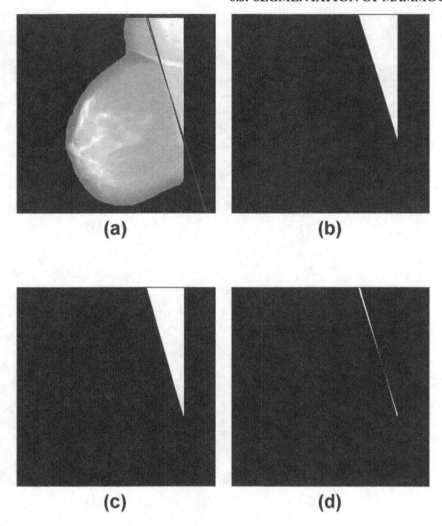

Figure 3.10: Example of accurate detection of the pectoral muscle region: (a) Image showing the straight line manually marked by a radiologist (in gray) and automatically detected by the method described (in black); (b) image of the pectoral muscle region segmented based on the manually marked straight line; (c) image of the pectoral muscle region segmented based on the straight line detected by the method described; (d) image with FP and FN regions obtained from the difference between the images in (b) and (c). Quantitative error values are FP = 1.2%, FN = 3.0%, FP and FN combined = 4.2%, and Hausdorff distance = 4.47 mm. Reproduced with kind permission from Springer Science+Business Media B. V. from S. K. Kinoshita, P. M. de Azevedo-Marques, R. R. Pereira Jr., J. A. H. Rodrigues, and R. M. Rangayyan, "Radon-domain detection of the nipple and the pectoral muscle in mammograms," Journal of Digital Imaging, 21(1): 37–49, March 2008. © Springer.

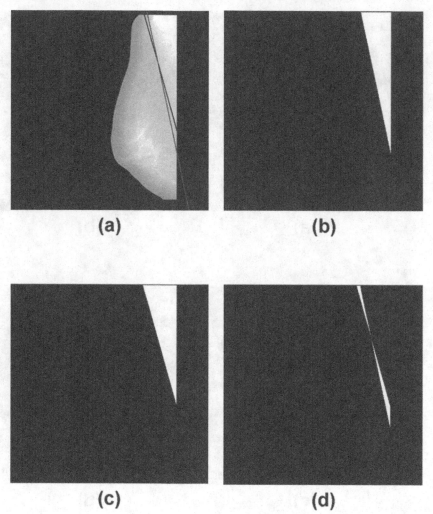

Figure 3.11: Example of acceptable detection of the pectoral muscle region: (a) Image showing the straight line manually marked by a radiologist (in gray) and automatically detected by the method described (in black); (b) image of the pectoral muscle region segmented based on the manually marked straight line; (c) image of the pectoral muscle region segmented based on the straight line detected by the method described; (d) image with FP and FN regions obtained from the difference between the images in (b) and (c). Quantitative error values are FP = 4.2%, FN = 9.9%, FP and FN combined = 14.1%, and Hausdorff distance = 5.31 mm. Reproduced with kind permission from Springer Science+Business Media B. V. from S. K. Kinoshita, P. M. de Azevedo-Marques, R. R. Pereira Jr., J. A. H. Rodrigues, and R. M. Rangayyan, "Radon-domain detection of the nipple and the pectoral muscle in mammograms," Journal of Digital Imaging, 21(1): 37–49, March 2008. © Springer.

The results of the method for the detection of the edge of the pectoral muscle show that the performance is related to the appearance of the edge of the pectoral muscle in the mammogram. Of the 376 images with results considered to be accurate and acceptable, 80% (301/376) had pectoral muscles with straight edges, 10.9% (41/376) had considerably curved edges, and 9.1% (34/376) had poorly defined edges with low contrast. Of the 164 images with results considered to be unacceptable, 14% (23/164) had pectoral muscles with straight edges, 37.2% (61/164) had considerably curved edges, and 48.8% (80/164) had poorly defined edges. The method provided better performance with images including pectoral muscles having straight edges and worse performance with other types of edges of the pectoral muscle region. See Ferrari et al. [77] for methods for detection of curvilinear edges of the pectoral muscle using Gabor filters.

3.3.5   DETECTION OF THE NIPPLE

Kinoshita et al. [83] developed a process for automatic detection of the position of the nipple in a mammographic image considering the following features of breast anatomy: the nipple is located on the breast surface; glandular tissues contain structures (such as lobes and ducts) that converge toward the nipple and appear as bright (radio-opaque) regions; and adipose and low-density tissues appear as darker regions in mammograms.

The morphological top-hat filter was applied to the mammogram to enhance the breast structures converging toward the nipple [91, 92]. The top-hat filter includes a closing operation in accordance with a structuring element, followed by subtraction of the filtered image from the original. A disk-shaped structuring element with a radius of 50 pixels was used. The filter enhances selected features based on the size and shape of the structuring element specified, and also darkens the background elements; see Figure 3.12. When processing MLO views, the pectoral muscle region was removed from the image before application of the top-hat filter.

The Radon transform was applied to the result of the top-hat filter obtained as described above. For MLO views, the angle intervals used were $45° + \alpha \leq \theta \leq 135° + \alpha$ for right-breast images and $-45° - \alpha \leq \theta \leq -135° - \alpha$ for left-breast images, where α is the direction of the pectoral muscle edge (straight-line approximation) and θ is the angle in the Radon domain. For CC views, the angle intervals were $45° \leq \theta \leq 135°$ and $-45° \leq \theta \leq -135°$ for right-breast and left-breast images, respectively. The point in the mammogram containing the highest number of converging lines is expected to be proximal to the nipple; see Figure 3.12. Given that the nipple is located on the breast surface (breast edge or boundary in the mammogram), its position was located as follows: the Radon transform of the gray-level mammogram (before application of the top-hat filter) was searched to locate the ray with the maximum intensity that crossed the point of convergence; this straight line was extended to the breast boundary to detect the nipple; see Figure 3.12.

Kinoshita et al. [83] tested the method for the detection of the nipple with 540 MLO and 540 CC mammographic images. Figure 3.13 shows four mammograms with different levels of error in the detection of the nipple. The absolute error between the detected position and that identified by the radiologist was, on the average, 7.4 mm over the 1,080 images processed. Table 3.1 shows the

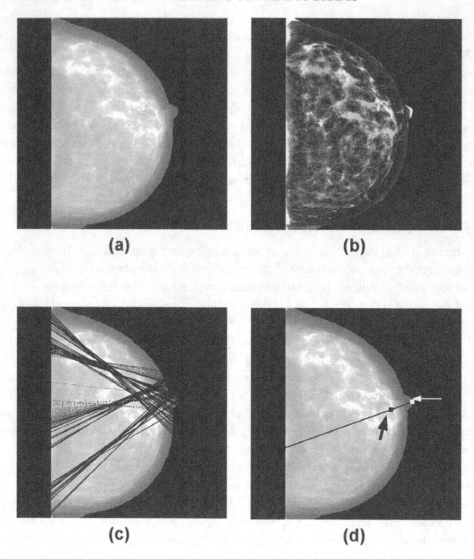

Figure 3.12: Illustration of the sequence of operations for the detection of the nipple: (a) original image (left breast, CC view); (b) result of the top-hat filter; (c) straight lines corresponding to the maximum values of the Radon transform in the directions $-45° \leq \theta \leq -135°$, where θ is the angle of the Radon transform (projection); (d) the white point (white arrow) indicates the nipple position detected and the black point (black arrow) represents the point of convergence. Reproduced with kind permission from Springer Science+Business Media B. V. from S. K. Kinoshita, P. M. de Azevedo-Marques, R. R. Pereira Jr., J. A. H. Rodrigues, and R. M. Rangayyan, "Radon-domain detection of the nipple and the pectoral muscle in mammograms," Journal of Digital Imaging, 21(1): 37–49, March 2008. © Springer.

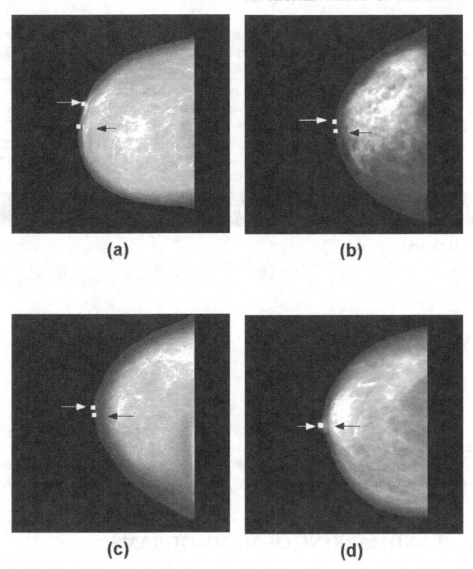

Figure 3.13: Examples of detection of the nipple: (a) error = 24.1 mm; (b) error = 9.8 mm; (c) error = 7.8 mm; and (d) error = 1.0 mm. Two points are shown on each mammogram, corresponding to the nipple positions detected automatically by the method described (black arrow) and manually marked by a radiologist (white arrow). Reproduced with kind permission from Springer Science+Business Media B. V. from S. K. Kinoshita, P. M. de Azevedo-Marques, R. R. Pereira Jr., J. A. H. Rodrigues, and R. M. Rangayyan, "Radon-domain detection of the nipple and the pectoral muscle in mammograms," Journal of Digital Imaging, 21(1): 37–49, March 2008. © Springer.

distribution of images by the range of error in mm. Out of the 1,080 images processed, the results were considered to be accurate (error $\leq$ 5 mm) in 585 images (54.2%); acceptable (5 mm < error $\leq$ 10 mm) in 271 images (25.1%); and not acceptable (error > 10 mm) in 224 images (20.7%).

Table 3.1: Distribution of images by the range of the absolute error in mm in the detection of the nipple as compared to the same identified by a radiologist. Reproduced with kind permission from Springer Science+Business Media B. V. from S. K. Kinoshita, P. M. de Azevedo-Marques, R. R. Pereira Jr., J. A. H. Rodrigues, and R. M. Rangayyan, "Radon-domain detection of the nipple and the pectoral muscle in mammograms," Journal of Digital Imaging, 21(1): 37–49, March 2008. © Springer.

Image type	$\leq$ 5 mm	5 to 10 mm	10 to 15 mm	15 to 20 mm	> 20 mm	Total
MLO right	146	52	27	16	29	270
MLO left	114	80	27	20	29	270
CC right	157	67	22	14	10	270
CC left	168	72	14	7	9	270
Total	585	271	90	57	77	1080

The results show that the performance of the method for the detection of the nipple is related to the quality of the image of the breast region after the initial preprocessing step. Of the 224 images with results considered to be not acceptable, 88.4% (198/224) images had been considered to be not acceptable or acceptable, and 11.6% (26/224) had been considered to be accurate after the initial preprocessing step.

3.4   LANDMARKING OF MAMMOGRAMS

Features in mammograms such as the edge of the pectoral muscle, the nipple, and the breast boundary are useful in further detailed analysis of mammograms. Initial identification of such features facilitates the design of algorithms for focused analysis of other parts of mammograms which may be identified and outlined using the initially detected features for reference. For this reason, the features mentioned above could be referred to as landmarks; the procedure of application of such information in further analysis is known as landmarking.

Figure 3.14 shows a schematic representation of the division of CC and MLO mammograms into two parts each. Because the pectoral muscle is usually not visible in a CC view, the vertical edge of the image connected to the breast region may be used to represent the chest wall. The nipple

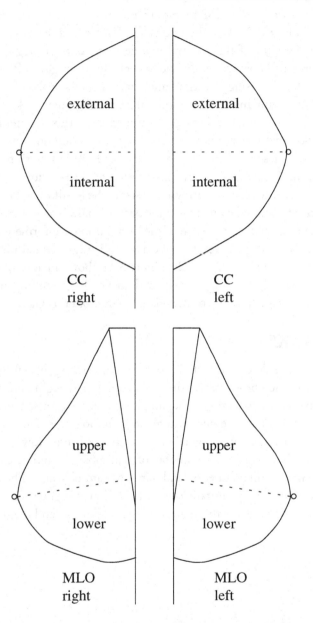

Figure 3.14: The breast region in each mammogram was divided in two subregions. Each CC view was divided into its external (lateral) and internal (medial) regions. Each MLO view was divided into its upper (superior) and lower (inferior) regions. Based on a similar figure in Kinoshita et al. [11].

location can be used to divide a CC mammogram into its external (lateral) and internal (medial) regions, as shown in Figure 3.14. In the case of an MLO view, the perpendicular to the straight-line approximation of the edge of the pectoral muscle that passes through the nipple can be used to divide the image into its upper (superior) and lower (inferior) regions. Partitioning of mammograms as above can be performed using the segmented breast regions or the fibroglandular disk. Figure 3.8 shows a CC mammogram and the corresponding fibroglandular disk segmented into its external and internal subregions. Segmentation of a mammogram in this manner facilitates focused analysis via feature extraction methods applied to the subregions, as demonstrated in the chapters to follow.

The landmarks described above were used in the present work to reformat mammographic images as follows. An image of a CC view was reformatted by positioning the automatically detected nipple location at the center of a new array and resizing the result to 1024×1024 pixels. An image of an MLO view was reformatted by removing the pectoral muscle region, rotating the image such that the straight line approximating the edge of the pectoral muscle formed the corresponding vertical edge of the image, centering the result at the automatically detected nipple position, and resizing the resulting array to 1024×1024 pixels. Figure 3.15 shows images of a CC view and an MLO view before and after reformatting as described above. The resulting images facilitate improved representation of the characteristics of the related mammograms via feature extraction.

3.5   REMARKS

In this chapter, we have described techniques for preprocessing of mammograms, segmentation of the breast region and the fibroglandular disk, detection of a straight-line approximation to the edge of the pectoral muscle, and detection of the nipple in mammograms. The Radon transform presents useful information related to oriented patterns in mammograms that can be related to the edge of the pectoral muscle and the parenchymal patterns that converge toward the nipple. The results of evaluation of the methods with a large number of mammograms indicate the success of the methods in the detection of important features or landmarks in mammograms. The following chapters present methods for the analysis and quantitative characterization of segmented regions that could be useful in CBIR or CAD algorithms for the analysis of mammograms and diagnosis of breast cancer.

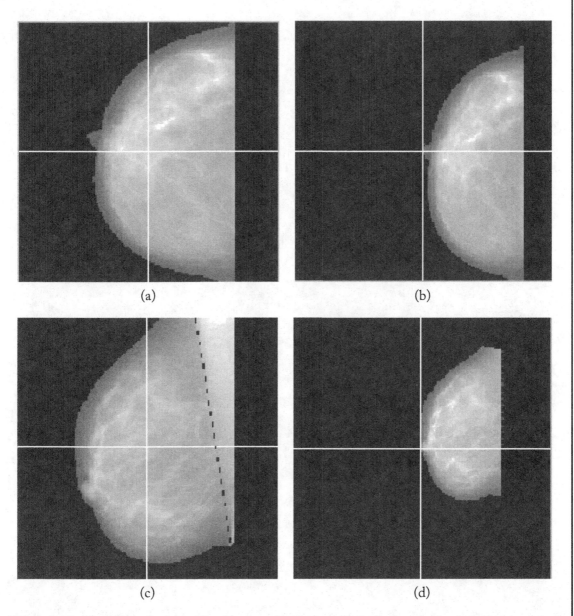

(a)   (b)

(c)   (d)

Figure 3.15: Examples of reformatting images using the automatically detected landmarks: (a) original CC view; (b) reformatted CC view; (c) original MLO view; and (d) reformatted MLO view. Figure courtesy of S. K. Kinoshita [93].

CHAPTER 4

Feature Extraction and Indexing of Mammograms

4.1 QUANTITATIVE REPRESENTATION OF MAMMOGRAPHIC FEATURES

Mammograms provide diagnostic information in several manners, including density or gray-level variations, texture, and shape. In a practical CAD or CBIR application, it is necessary to reduce the pixel-based information present in mammograms to a small number of measures or features that represent diagnostically useful information in a compact and efficient manner. The features should also facilitate clear distinction or discrimination between various categories or classes of images, as well as aid in the detection and analysis of abnormal entities such as masses, tumors, and calcifications. If an appropriate set of features is obtained, it becomes possible to represent a given image in terms of the set or vector of features for further analysis instead of the pixel-based image data. The present chapter provides descriptions of several methods to derive features to represent the gray-level variations, texture, and shape details present in mammographic images.

When a number of features are derived from an image based on prior knowledge of the expected characteristics and their variation between various classes of images, difficulties may be encountered in further analysis if the dimension of the feature vector is large. Furthermore, some of the features may be correlated with others, which can cause inefficiency in further analysis. For these reasons, it is necessary to select an efficient subset of the features. Methods for feature extraction and selection are described in the present chapter with the aim of efficient representation of mammograms for CAD and CBIR.

4.2 STATISTICAL ANALYSIS OF GRAY-LEVEL VARIATION

The distribution of gray levels in an image over the available range is represented by its histogram; see Figure 2.5. The histogram provides quantitative information on the probability of occurrence of each gray level in the image. It is desirable to express, in a few measures, the manner in which the values of a histogram or PDF vary over the available range. Such measures may then be used as features to represent the image.

Entropy is a statistical measure of information that is commonly used for this purpose [3, 94, 99–101]. The pixels in an image may be considered to be symbols produced by a discrete information source with the gray levels as its states. Consider the occurrence of L gray levels in an image, with the probability of occurrence of the lth gray level being $p(l), l = 0, 1, 2, \ldots, L - 1$. Let us treat the gray level of a pixel as a random variable. A measure of information conveyed by an event, in the present case a pixel or a gray level, may be related to the statistical uncertainty of the event giving rise to the information, rather than the semantic or structural content of the signal or image. Entropy is defined as the expected value of information contained in each possible level:

$$H = - \sum_{l=0}^{L-1} p(l) \, \log_2 [p(l)]. \tag{4.1}$$

The unit of entropy is bits (b). Because the gray levels are considered as individual entities in this definition, that is, no neighboring elements are taken into account, the result is known as the zeroth-order entropy [102].

Measures of texture may be derived from the moments of the gray-level PDF of the given image [3]. The kth central moment of the PDF $p(l)$ is defined as

$$\mu_k = \sum_{l=0}^{L-1} (l - \mu_f)^k \, p(l), \tag{4.2}$$

where μ_f is the mean gray level of the image f, given by

$$\mu_f = \sum_{l=0}^{L-1} l \, p(l). \tag{4.3}$$

The second central moment, which is the variance of the gray levels, given by

$$\sigma_f^2 = \mu_2 = \sum_{l=0}^{L-1} (l - \mu_f)^2 \, p(l), \tag{4.4}$$

can serve as a measure of inhomogeneity. The square root of variance is known as standard deviation (σ).

The normalized third and fourth moments, known as skewness and kurtosis, respectively, and defined as

$$\text{skewness} = \frac{\mu_3}{\mu_2^{3/2}}, \tag{4.5}$$

and

$$\text{kurtosis} = \frac{\mu_4}{\mu_2^2}, \tag{4.6}$$

indicate the asymmetry and uniformity (or lack thereof) of the PDF. High-order moments are affected by noise or errors in the PDF, and may not be reliable features. The moments of the PDF can serve as basic representatives of gray-level variation.

Byng et al. [53] computed the skewness of the histograms of sections of size 24 × 24 pixels (3.12 × 3.12 mm) obtained from mammograms. An average skewness measure was computed for each image by averaging over all the section-based skewness measures of the image. Mammograms of breasts with increased fibroglandular density were observed to have histograms skewed toward higher density (see Figure 2.5), resulting in negative skewness. Mammograms of fatty breasts tended to have positive skewness. Skewness was found to be useful in predicting the risk of development of breast cancer.

4.3   STATISTICAL ANALYSIS OF TEXTURE

Notions of texture in an image are related to the spatial arrangement of intensity or gray levels in the image. Differences have been observed in texture between benign masses and malignant tumors, with the former being mostly homogeneous and the latter showing heterogeneous texture [10, 27, 34]. Several studies have applied various measures of texture to discriminate between benign masses and malignant tumors [10, 66, 103–105]. The most commonly used measures of texture are those proposed by Haralick et al. [15], which are briefly described below.

The measures of texture proposed by Haralick et al. [15] are statistical parameters derived from the GLCM of the given image. The GLCM is a square matrix with its number of rows and columns equal to the number of gray levels in the image. For an image with B bpp, there will be $L = 2^B$ gray levels; the GLCM is of size $L \times L$. For an image quantized to 8 bpp, there will be 256 gray levels; the GLCM is of size 256 × 256. The GLCM of an image provides normalized values of the numbers of occurrences of all combinations of pixel pairs, separated by a specified distance, d, and angle, θ, that occur in the image. GLCMs are commonly formed for unit pixel distance and the four angles of 0°, 45°, 90°, and 135°.

Table 4.1 shows the GLCM for the image in Figure 4.1 with eight gray levels (3 bpp) by considering pairs of pixels with the second pixel immediately next to (the right of) the first. For example, the pair of gray levels [1 2] occurs three times in the image. Because neighboring pixels in natural images tend to have nearly the same values, GLCMs usually have large values along and around the main diagonal, and small values away from the diagonal.

Digital or digitized mammograms are usually acquired at the resolution of about 50μm per pixel, with 4096 gray levels represented using 12 bpp. The GLCM of a 12-bpp image, being of size 4096 × 4096, would be excessively large for practical computation. Furthermore, most pairs of gray levels would occur with low or negligible rates of incidence that may not permit the derivation of reliable statistics. Therefore, it is advantageous to reduce the image to 256 gray levels, that is, to use 8 bpp. The images in the datasets used in the present work were requantized at 8 bpp. Lee et al. [106] conducted a study on the effect of the number of gray levels used on texture analysis of mammograms.

Table 4.1: Gray-level cooccurrence matrix for the image in Figure 4.1, with the second pixel immediately next to (the right of) the first. Pixels in the last column were not processed as current pixels. The GLCM has not been normalized.

Current Pixel	Next Pixel							
	0	1	2	3	4	5	6	7
0	0	0	1	0	1	0	0	0
1	0	0	3	1	0	5	0	0
2	0	0	1	12	12	5	1	1
3	0	1	10	23	23	6	4	1
4	0	3	9	18	26	14	4	2
5	0	3	6	6	11	7	2	1
6	0	0	3	5	3	0	0	0
7	2	1	0	0	1	1	0	1

The GLCM is computed as

$$p_{(d, \theta)}(l_1, l_2) = \frac{n_{l_1 l_2}}{\sum_{l_1=0}^{L-1} \sum_{l_2=0}^{L-1} n_{l_1 l_2}}, \tag{4.7}$$

where L is the number of gray levels in the image; l_1 and l_2, with each variable in the range $0, 1, 2, \ldots, L - 1$, represent a pair of gray levels; $n_{l_1 l_2}$ is the number of occurrences of the pair of gray levels l_1 and l_2; and $p_{(d, \theta)}(l_1, l_2)$ is the probability of occurrence of the pair of gray levels l_1 and l_2 at the specified spatial distance d and angle θ. A normalized GLCM as above may be treated as a 2D PDF.

A GLCM may be computed for each pair of d and θ values of interest to analyze the dependence of texture upon spatial distance and angle, or a single GLCM, $p(l_1, l_2)$, may be computed as the average of the GLCMs for all angles (usually, $\theta = 0°, 45°, 90°, 135°$) at a given distance d if the dependence of texture upon angle is not of interest. The distance d should be chosen with consideration of the size (or resolution) of the objects of interest or scale of variation present in the

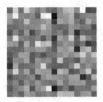

```
2  6  3  6  4  3  4  3  2  4  5  5  4  7  1  2
4  7  4  3  4  2  7  7  0  4  4  5  2  3  4  4
4  3  3  5  4  4  4  5  1  5  4  5  3  2  5  4
3  3  3  4  4  4  5  4  3  3  3  4  4  3  3  2
2  4  2  4  2  3  4  4  3  4  3  4  1  3  2  2
3  6  4  3  3  4  2  4  1  5  4  6  2  5  4  3
4  4  2  5  3  7  5  6  2  4  4  4  2  3  1  2
3  3  3  4  4  4  3  2  5  4  5  1  5  5  2
2  3  5  5  2  3  2  3  3  4  5  4  5  6  4  5
2  4  4  5  2  4  4  4  3  6  3  3  2  4  4  4
1  5  3  3  3  4  6  3  3  2  3  4  6  3  4  3
3  6  3  5  3  5  4  4  1  5  5  2  3  4  6  2
2  3  3  3  4  2  4  5  4  4  4  5  5  1  2  5
3  4  2  4  3  4  3  3  3  5  3  4  4  4  3  3
2  3  4  5  2  4  5  7  0  2  3  2  3  3  3  4
3  3  2  4  4  3  3  4  3  5  5  5  3  4  4  2
```

Figure 4.1: A 16×16-pixel part of a mammogram quantized to 3 bpp, shown as an image and as a 2D array of pixel values.

image. The choice of d determines the resolution at which the texture is being analyzed. Rangayyan et al. [107] presented a study analyzing the effect of spatial resolution or pixel size on GLCM-based texture features in the classification of breast masses in mammograms; see also Cabral and Rangayyan [108].

Fourteen statistical measures were derived from GLCMs by Haralick et al. [15], which are briefly described in the following paragraphs.

Angular second moment: A measure of homogeneity or uniformity in the image. A homogeneous image has similar pixel intensity throughout; thus, the GLCM of the image will have a small number of high values concentrated along the diagonal. On the other hand, the GLCM of an inhomogeneous image will have small values spread over a large number of entries of its matrix. The angular second moment is defined as

$$F_1 = \sum_{l_1=0}^{L-1} \sum_{l_2=0}^{L-1} p^2(l_1, l_2).$$

(4.8)

Contrast: A measure of the amount of local variations present in the image. Contrast is defined as

$$F_2 = \sum_{k=0}^{L-1} k^2 \underbrace{\sum_{l_1=0}^{L-1} \sum_{l_2=0}^{L-1} p(l_1, l_2)}_{|l_1-l_2|=k}.$$

(4.9)

Correlation: A measure of linear dependencies of gray levels in the image (that is, how a pixel is correlated to its neighborhood). The correlation measure is defined as

$$F_3 = \frac{1}{\sigma_x \, \sigma_y} \left[\sum_{l_1=0}^{L-1} \sum_{l_2=0}^{L-1} l_1 \, l_2 \, p(l_1, l_2) - \mu_x \, \mu_y \right],$$

(4.10)

where μ_x and μ_y are the means, and σ_x and σ_y are the standard deviation values of p_x and p_y, respectively. The marginal probabilities, p_x and p_y, are defined as

$$p_x(l_1) = \sum_{l_2=0}^{L-1} p(l_1, l_2),$$

(4.11)

$$p_y(l_2) = \sum_{l_1=0}^{L-1} p(l_1, l_2).$$

(4.12)

Sum of squares: A measure of the gray-level variance in the image.

$$F_4 = \sum_{l_1=0}^{L-1} \sum_{l_2=0}^{L-1} (l_1 - \mu_f)^2 \, p(l_1, l_2),$$

(4.13)

where μ_f is the mean gray level of the image.

Inverse difference moment: A measure of local uniformity present in the image. The inverse difference moment is defined as

$$F_5 = \sum_{l_1=0}^{L-1} \sum_{l_2=0}^{L-1} \frac{1}{1 + (l_1 - l_2)^2} \, p(l_1, l_2).$$

(4.14)

This feature may be seen as the inverse of the contrast feature defined in Equation 4.9. Thus, the inverse difference moment is high for images having low contrast and low for images having high contrast.

Sum average: The sum average is defined as

$$F_6 = \sum_{k=0}^{2(L-1)} k \; p_{x+y}(k) \, , \tag{4.15}$$

where p_{x+y} is given by

$$p_{x+y}(k) = \underbrace{\sum_{l_1=0}^{L-1} \sum_{l_2=0}^{L-1}}_{l_1+l_2=k} p(l_1, l_2) \, . \tag{4.16}$$

Sum variance: The sum variance is defined as

$$F_7 = \sum_{k=0}^{2(L-1)} (k - F_6)^2 \; p_{x+y}(k) \, . \tag{4.17}$$

Sum entropy: The sum entropy is defined as

$$F_8 = - \sum_{k=0}^{2(L-1)} p_{x+y}(k) \; \log_2 \left[p_{x+y}(k) \right] \, . \tag{4.18}$$

Entropy: A measure of the nonuniformity or complexity of the texture of the image. The entropy is defined as

$$F_9 = - \sum_{l_1=0}^{L-1} \sum_{l_2=0}^{L-1} p(l_1, l_2) \; \log_2 \left[p(l_1, l_2) \right] \, . \tag{4.19}$$

Difference variance: The difference variance is computed, in a manner similar to the sum variance, for the p_{x-y} matrix, as

$$F_{10} = \sum_{k=0}^{2(L-1)} \left(k - \sum_{k=0}^{2(L-1)} k \; p_{x-y}(k) \right)^2 p_{x-y}(k) \, , \tag{4.20}$$

where p_{x-y} is given by

$$p_{x-y}(k) = \underbrace{\sum_{l_1=0}^{L-1} \sum_{l_2=0}^{L-1} p(l_1, l_2)}_{|l_1-l_2|=k}. \tag{4.21}$$

Difference entropy: The difference entropy is defined as

$$F_{11} = -\sum_{k=0}^{L-1} p_{x-y}(k) \ \log_2 \left[p_{x-y}(k) \right]. \tag{4.22}$$

Two information measures of correlation: The information measures of correlation are defined as

$$F_{12} = \frac{H_{xy} - H_{xy1}}{\max\{H_x, H_y\}}, \tag{4.23}$$

and

$$F_{13} = \left\{ 1 - \exp[-2\left(H_{xy2} - H_{xy}\right)] \right\}^{\frac{1}{2}}, \tag{4.24}$$

where $H_{xy} = F_9$; H_x and H_y are the entropies of p_x and p_y, respectively;

$$H_{xy1} = -\sum_{l_1=0}^{L-1} \sum_{l_2=0}^{L-1} p(l_1, l_2) \ \log_2 \left[p_x(l_1) \ p_y(l_2) \right], \tag{4.25}$$

and

$$H_{xy2} = -\sum_{l_1=0}^{L-1} \sum_{l_2=0}^{L-1} p_x(l_1) \ p_y(l_2) \ \log_2 \left[p_x(l_1) \ p_y(l_2) \right]. \tag{4.26}$$

Maximum correlation coefficient: The maximum correlation coefficient is defined as

$$F_{14} = (Second\ largest\ eigenvalue\ of\ Q)^{1/2}, \tag{4.27}$$

where Q is computed as

$$Q(l_1, l_2) = \sum_{k=0}^{L-1} \frac{p(l_1, k) \ p(l_2, k)}{p_x(l_1) \ p_y(k)}. \tag{4.28}$$

Further details of the characteristics of the 14 statistical measures of texture based upon the GLCM are given by Haralick et al. [15]. These measures represent characteristics of the spatial distribution of gray levels in the image. Texture analysis, using some or all of the 14 Haralick's texture features, is a popular approach for the analysis and classification of many medical images, including breast masses and tumors seen in mammograms [3]. Therefore, in the present work, Haralick's texture measures were chosen for the analysis of the texture in mammograms.

4.4   GRANULOMETRIC ANALYSIS

Granulometry provides measures that characterize the distribution of the size of objects in an image [109]. To obtain granulometric features, the morphological opening filter was applied to the segmented fibroglandular disks. The structuring element was defined as a circle with radius varying between zero and 60 pixels (in steps of five pixels). For each structuring element, a difference image was computed between the original and the filtered versions. A histogram was created to represent the number of nonzero (fibroglandular tissue) pixels in the difference images against the size of the structuring element. The 12 values of the histogram were used as features.

Figure 4.2 shows examples of histograms obtained as above for two mammograms. In these examples, the structuring element was defined as a circle with radius varying between zero and 60 pixels, in steps of one pixel. Peaks in the histograms are related to the sizes of scattered dense regions present in the mammograms.

4.5   ANALYSIS OF SHAPE

The goal of shape analysis in the present work is to find other images with the same size and shape of the breast (in terms of being elongated or rounded in shape and rough in texture) as the query image in the retrieval process. Because a mammogram is a 2D representation of the 3D interaction between X rays and breast tissue, some bias and error can be introduced in density evaluation. For instance, women with large breasts may have a substantial amount of glandular breast tissue, and yet have an appearance of low breast density in their mammograms. On the other hand, women with small breasts and low amounts of glandular tissue may have the appearance of higher density in their mammograms. Therefore, shape and size features can have an important complementary role in density evaluation.

4.5.1   MORPHOMETRIC AND SHAPE FACTORS

In the present study, three shape and size features are used: area, ratio of diameters, and compactness. The area A of each segmented breast region was estimated in pixels and normalized with respect to the total number of pixels in the image. A correction factor was applied to take into account the two sizes of mammograms in the database and the related pixel sizes.

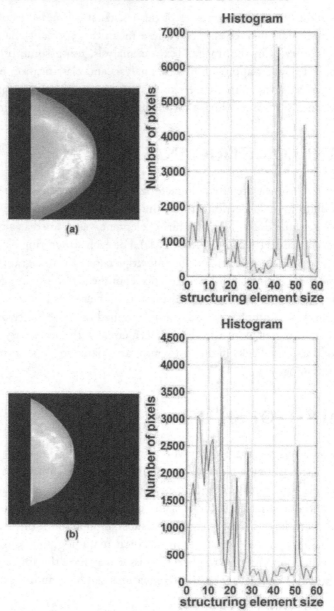

Figure 4.2: Illustration of the procedure to compute granulometric features using two mammograms. Reproduced with kind permission from Springer Science+Business Media B. V. from S. K. Kinoshita, P. M. de Azevedo-Marques, R. R. Pereira Jr., J. A. H. Rodrigues, and R. M. Rangayyan, "Content-based retrieval of mammograms using visual features related to breast density patterns," Journal of Digital Imaging, 20(2): 172–190, June 2007. © Springer.

The ratio of diameters DR was defined as the ratio

$$DR = \frac{D_y}{D_x}. \tag{4.29}$$

The distance D_y was measured as the distance from the nipple to the chest wall in CC views, and as the perpendicular distance from the pectoral muscle edge to the nipple in the case of MLO views (see Figure 4.3). The distance D_x was defined as the length of the perpendicular bisector of the line related to D_y (within the breast region), as shown in Figure 4.3. The feature DR characterizes the elongation of the breast.

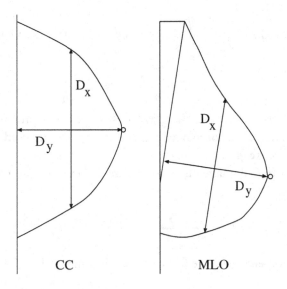

Figure 4.3: Illustration of the distances D_x and D_y used to derive DR, the ratio of diameters. Based on a similar figure in Kinoshita et al. [11].

Compactness is a commonly used measure of shape roughness, defined as the ratio of the squared perimeter, P^2, to the area, A, of a region or contour [3]. The normalized version of compactness, given by

$$C = 1 - \frac{2\pi A}{P^2}, \tag{4.30}$$

was used to derive a shape factor for the binarized breast region.

See Rangayyan [3] for details on several other shape factors useful in biomedical image analysis.

4.5.2 SHAPE ANALYSIS USING MOMENTS

Statistical moments of gray-level PDFs, as described in Section 4.2, can be used as features to represent images in a number of applications; the same concepts can be extended to the analysis of shape in images and contours [3, 91, 103, 110–112].

Given a 2D continuous image $f(x, y)$, the regular moments m_{pq} of order $(p + q)$ are defined as [91, 112]:

$$m_{pq} = \int_{-\infty}^{+\infty} \int_{-\infty}^{+\infty} x^p y^q f(x, y) \, dx \, dy, \qquad (4.31)$$

for $p, q = 0, 1, 2, \ldots$. The central moments are defined with respect to the centroid of the image as

$$\mu_{pq} = \int_{-\infty}^{+\infty} \int_{-\infty}^{+\infty} (x - \overline{x})^p (y - \overline{y})^q f(x, y) \, dx \, dy, \qquad (4.32)$$

where

$$\overline{x} = \frac{m_{10}}{m_{00}}, \qquad \overline{y} = \frac{m_{01}}{m_{00}}. \qquad (4.33)$$

In the definitions given above, the gray levels of the pixels provide weights for the moments. If moments are computed for a contour, only the contour pixels are used with weights equal to unity; the internal pixels have weights of zero.

For an $M \times N$ digital image, the integrals are replaced by summations:

$$\mu_{pq} = \sum_{m=0}^{M-1} \sum_{n=0}^{N-1} (m - \overline{x})^p (n - \overline{y})^q f(m, n). \qquad (4.34)$$

The central moments have the following relationships [91]:

$$
\begin{aligned}
\mu_{00} &= m_{00} = \mu, & (4.35) \\
\mu_{10} &= \mu_{01} = 0, & (4.36) \\
\mu_{20} &= m_{20} - \mu\overline{x}^2, & (4.37) \\
\mu_{11} &= m_{11} - \mu\overline{x}\,\overline{y}, & (4.38) \\
\mu_{02} &= m_{02} - \mu\overline{y}^2, & (4.39) \\
\mu_{30} &= m_{30} - 3m_{20}\overline{x} + 2\mu\overline{x}^3, & (4.40) \\
\mu_{21} &= m_{21} - m_{20}\overline{y} - 2m_{11}\overline{x} + 2\mu\overline{x}^2\overline{y}, & (4.41) \\
\mu_{12} &= m_{12} - m_{02}\overline{x} - 2m_{11}\overline{y} + 2\mu\overline{x}\,\overline{y}^2, & (4.42) \\
\mu_{03} &= m_{03} - 3m_{02}\overline{y} + 2\mu\overline{y}^3. & (4.43)
\end{aligned}
$$

Normalization with respect to the size of the image is achieved by dividing each of the moments by μ_{00}^{γ}, where $\gamma = \frac{p+q}{2} + 1$, to obtain the normalized moments as [91]

$$\nu_{pq} = \frac{\mu_{pq}}{\mu_{00}^{\gamma}}. \qquad (4.44)$$

Hu [112] defined a set of seven shape factors that are functions of the second-order and third-order central moments as follows:

$$M_1 = v_{20} + v_{02}, \qquad (4.45)$$

$$M_2 = (v_{20} - v_{02})^2 + 4v_{11}^2, \qquad (4.46)$$

$$M_3 = (v_{30} - 3v_{12})^2 + (3v_{21} - v_{03})^2, \qquad (4.47)$$

$$M_4 = (v_{30} + v_{12})^2 + (v_{21} + v_{03})^2, \qquad (4.48)$$

$$M_5 = (v_{30} - 3v_{12})(v_{30} + v_{12})[(v_{30} + v_{12})^2 - 3(v_{21} + v_{03})^2] \\ + (3v_{21} - v_{03})(v_{21} + v_{03})[3(v_{30} + v_{12})^2 - (v_{21} + v_{03})^2], \qquad (4.49)$$

$$M_6 = (v_{20} - v_{02})[(v_{30} + v_{12})^2 - (v_{21} + v_{03})^2] \\ + 4v_{11}(v_{30} + v_{12})(v_{21} + v_{03}), \qquad (4.50)$$

$$M_7 = (3v_{21} - v_{03})(v_{30} + v_{12})[(v_{30} + v_{12})^2 - 3(v_{21} + v_{03})^2] \\ - (v_{30} - 3v_{12})(v_{21} + v_{03})[3(v_{30} + v_{12})^2 - (v_{21} + v_{03})^2]. \qquad (4.51)$$

The shape factors M_1 through M_7 are invariant to shift, scaling, and rotation, and have been found to be useful for pattern analysis. Rangayyan et al. [103] computed several versions of the factors M_1 through M_7 for 54 breast masses and tumors, using the mass ROIs with and without their gray levels, as well as the contours of the masses with and without their gray levels. The features provided benign-versus-malignant classification accuracies in the range $56 - 75\%$.

In the present work, the seven moment-based features were computed for each segmented subregion, as shown in Figure 3.14.

4.6 FEATURE EXTRACTION, SELECTION, AND INDEXING OF IMAGES

4.6.1 FEATURE EXTRACTION

In the work by Kinoshita et al. [11], several morphological, textural, and statistical measures were extracted from various segmented mammographic images for CBIR. Table 4.2 lists the types and numbers of features extracted. The three shape factors described in Section 4.5.1 were computed from the binarized breast region [as shown in Figure 3.4 (c)] for each mammogram. For each subregion representing the internal or external region of a CC view or the upper or lower region of an MLO view, the 14 texture features described in Section 4.3, the seven moment-based features described in Section 4.5.2, and the five statistical measures (entropy, mean, standard deviation, skewness, and kurtosis) based on gray-level histograms, as described in Section 4.2, were computed. The two 1D functions derived from the Radon transform of each subregion (see Figure 3.8) were normalized so as to be treated as PDFs, and the five statistical measures, as described in Section 4.2 and mentioned above, were computed. A set of 12 measures were obtained from the histogram resulting from granulometric analysis of each segmented fibroglandular disk (see Section 4.4). In this manner, 87 features were obtained for each mammogram. Each feature was normalized to the range [0, 1] by subtracting its minimum and dividing by the difference between its maximum and minimum over the dataset.

Table 4.2: Number of features extracted in each group per ROI and mammogram. Texture: Haralick's texture features. Histogram: Statistics of gray-level histogram. Moments: Hu's moment-based features. Radon: Radon-domain features. Shape: Shape features. Granulometric: Granulometric features. *: internal or external region for CC views; upper or lower region for MLO views. Reproduced with kind permission from Springer Science+Business Media B. V. from S. K. Kinoshita, P. M. de Azevedo-Marques, R. R. Pereira Jr., J. A. H. Rodrigues, and R. M. Rangayyan, "Content-based retrieval of mammograms using visual features related to breast density patterns," Journal of Digital Imaging, 20(2): 172–190, June 2007. © Springer.

Features	ROI	Per ROI	Per mammogram
Shape	Binary breast region	3	3
Texture	Subregion*	14	28
Moments	Subregion*	7	14
Histogram	Subregion*	5	10
Radon	Subregion*	10	20
Granulometric	Fibroglandular disk	12	12

4.6.2   FEATURE SELECTION

In order to facilitate efficient pattern classification, CAD, or CBIR, it is important to prepare a compact set of features with no redundancy or mutual correlation and to keep the dimension of the feature set or vector as small as possible. Several feature selection methods are available for this purpose, including sequential forward selection, sequential backward selection, stepwise logistic regression, and genetic programming [113–115]. Another approach to reduce the dimension of a feature vector is to perform principal component analysis (PCA) and select a compact set of principal components. The PCA method, which is related to the Karhunen–Loève transform [3, 91, 99], was used in the present work and is described briefly in the following paragraphs.

Consider a set of features represented as a vector $\mathbf{f}$ of size P. The vector $\mathbf{f}$ may be represented without error by a deterministic linear transformation of the form

$$\mathbf{f} = \mathbf{A}\,\mathbf{g} = \sum_{m=1}^{P} g_m\,\mathbf{A}_m, \tag{4.52}$$

$$\mathbf{A} = \begin{bmatrix} \mathbf{A}_1; & \mathbf{A}_2; \cdots ; \mathbf{A}_P \end{bmatrix}, \tag{4.53}$$

where $|\mathbf{A}| \neq 0$, and $\mathbf{A}_m$ are $1 \times P$ row vectors that make up the $P \times P$ matrix $\mathbf{A}$. The matrix $\mathbf{A}$ needs to be formulated such that the vector $\mathbf{g}$ leads to a reduced and efficient representation of the original vector $\mathbf{f}$.

The matrix $\mathbf{A}$ may be considered to be made up of P linearly independent row vectors that span the P-dimensional space containing $\mathbf{f}$. Let $\mathbf{A}$ be orthonormal, that is,

$$\mathbf{A}_m^T\,\mathbf{A}_n \;=\; \begin{cases} 1 & m = n \\ 0 & m \neq n \end{cases}. \tag{4.54}$$

Then

$$\mathbf{A}^T\mathbf{A} = \mathbf{I} \text{ or } \mathbf{A}^{-1} = \mathbf{A}^T. \tag{4.55}$$

The row vectors of $\mathbf{A}$ may be considered to form the set of orthonormal basis vectors of a linear transformation. The inverse relationship is given by

$$\mathbf{g} = \mathbf{A}^T\,\mathbf{f} = \sum_{m=1}^{P} \mathbf{A}_m^T\,f_m. \tag{4.56}$$

In the formulation given above, each component of $\mathbf{g}$ contributes to the representation of $\mathbf{f}$. With the formulation of $\mathbf{A}$ as a reversible linear transformation, $\mathbf{g}$ provides a complete or lossless representation of $\mathbf{f}$ if all of its P elements are used.

In the interest of compact and efficient representation of the original set of features via the extraction of the most significant information contained, we may choose to use only $Q < P$ components of $\mathbf{g}$. The omitted components of $\mathbf{g}$ may be replaced with other values b_m, $m = Q + 1, Q + 2, \cdots, P$. Then we have an approximate representation of $\mathbf{f}$, as

$$\tilde{\mathbf{f}} = \sum_{m=1}^{Q} g_m\,\mathbf{A}_m + \sum_{m=Q+1}^{P} b_m\,\mathbf{A}_m. \tag{4.57}$$

The error in the approximate representation is

$$\boldsymbol{\varepsilon} = \mathbf{f} - \tilde{\mathbf{f}} = \sum_{m=Q+1}^{P} (g_m - b_m)\,\mathbf{A}_m. \tag{4.58}$$

The MSE is

$$\overline{\varepsilon^2} = E[\boldsymbol{\varepsilon}^T \boldsymbol{\varepsilon}]$$

$$= E[\sum_{m=Q+1}^{P} \sum_{n=Q+1}^{P} (g_m - b_m)(g_n - b_n) \mathbf{A}_m^T \mathbf{A}_n] \qquad (4.59)$$

$$= \sum_{m=Q+1}^{P} E[(g_m - b_m)^2].$$

The last step above is based on the orthonormality of $\mathbf{A}$.

Taking the derivative of the MSE with respect to b_m and setting the result to zero, we get

$$\frac{\partial \overline{\varepsilon^2}}{\partial b_m} = -2 E[(g_m - b_m)] = 0. \qquad (4.60)$$

The optimal or minimum-MSE (MMSE) choice for b_m is given by

$$b_m = E[g_m] = \overline{g}_m = \mathbf{A}_m^T E[\mathbf{f}], \quad m = Q + 1, Q + 2, \ldots, P; \qquad (4.61)$$

that is, the omitted components are replaced by their means computed for a given population of feature vectors. The MMSE is given by

$$\overline{\varepsilon^2}_{\min} = \sum_{m=Q+1}^{P} E[(g_m - \overline{g}_m)^2]$$

$$= \sum_{m=Q+1}^{P} E[\mathbf{A}_m^T (\mathbf{f} - \overline{\mathbf{f}})(\mathbf{f} - \overline{\mathbf{f}})^T \mathbf{A}_m] \qquad (4.62)$$

$$= \sum_{m=Q+1}^{P} \mathbf{A}_m^T \boldsymbol{\sigma}_f \mathbf{A}_m,$$

where $\boldsymbol{\sigma}_f$ is the covariance matrix of $\mathbf{f}$.

If the basis vectors $\mathbf{A}_m$ are selected as the eigenvectors of $\boldsymbol{\sigma}_f$, that is,

$$\boldsymbol{\sigma}_f \mathbf{A}_m = \lambda_m \mathbf{A}_m, \qquad (4.63)$$

and

$$\lambda_m = \mathbf{A}_m^T \boldsymbol{\sigma}_f \mathbf{A}_m \qquad (4.64)$$

because $\mathbf{A}_m^T \mathbf{A}_m = 1$, where λ_m are the corresponding eigenvalues, then we have

$$\overline{\varepsilon^2}_{\min} = \sum_{m=Q+1}^{P} \lambda_m. \qquad (4.65)$$

Therefore, the MSE may be minimized by ordering the eigenvectors (the rows of $\mathbf{A}$) such that the corresponding eigenvalues are arranged in decreasing order, that is, $\lambda_1 > \lambda_2 > \cdots > \lambda_P$. Then if a component g_m of $\mathbf{g}$ is replaced by $b_m = \overline{g}_m$, the MSE increases by λ_m. By replacing the components of **g** corresponding to the eigenvalues at the lower end of the list, the MSE is kept at its minimum for a chosen number of components Q.

From the formulation and properties described above, it is evident that the components of **g** are mutually uncorrelated:

$$\boldsymbol{\sigma}_g = \mathbf{A}^T \boldsymbol{\sigma}_f \mathbf{A} = \begin{bmatrix} \lambda_1 & & & \\ & \lambda_2 & & \\ & & \ddots & \\ & & & \lambda_P \end{bmatrix} = \boldsymbol{\Lambda}, \quad (4.66)$$

where $\boldsymbol{\Lambda}$ is a diagonal matrix with the eigenvalues λ_m placed along its diagonal. Because the eigenvalues λ_m are equal to the variances of g_m, a selection of the larger eigenvalues implies the selection of the transformed components with the higher variance or information content across the ensemble of the feature vectors considered. It should be noted that the features resulting from PCA are transformed versions of the original set of features and not a selected subset thereof.

In the present work, the transformed components of each set of features evaluated were selected so as to contain at least 95% of the total variance. PCA was applied to various subsets of the total of 87 features computed for each mammogram. Table 4.3 shows the effect of PCA in terms of reduction of the dimension of each feature subset. Considering the full set of 87 features, PCA resulted in a reduced set of 18 components, indicating more than four-fold reduction in the dimension of the feature vector.

4.6.3   INDEXING OF IMAGES FOR CBIR

The reduced set of components obtained via PCA of a selected set of features may now be used for quantitative representation of a mammographic image. A mammographic image, initially represented with millions of pixels, is effectively characterized by an ordered set or vector of a small number of values, as indicated in Table 4.3, derived from quantitative measures or features computed from the image. Such a representation may be referred to as indexing of the image, not in terms of the patient's identification number or a label, but in terms of numerical attributes that represent the content of the image in a manner that facilitates objective analysis via algorithms for CBIR or CAD.

4.7   REMARKS

In this chapter, we have described several methods to extract quantitative measures or features from mammograms or segmented regions thereof. We have shown how various radiological attributes of mammograms may be represented quantitatively using statistical and morphometric measures. We have also described the method of PCA as an approach to reduce the dimension of an initial set of features to facilitate efficient representation and indexing of images for CBIR and CAD. Results of application of the methods described in the present and preceding chapters to a large database of mammograms are presented in the following chapter.

Table 4.3: Number of features extracted in each group and dimension of features reduced by PCA. Texture: Haralick's texture features. Histogram: Statistics of gray-level histogram. Moments: Hu's moment-based features. Radon: Radon-domain features. Shape: Shape features. Granulometric: Granulometric features. Reproduced with kind permission from Springer Science+Business Media B. V. from S. K. Kinoshita, P. M. de Azevedo-Marques, R. R. Pereira Jr., J. A. H. Rodrigues, and R. M. Rangayyan, "Content-based retrieval of mammograms using visual features related to breast density patterns," Journal of Digital Imaging, 20(2): 172–190, June 2007. © Springer.

Features	Number	After PCA
Shape	3	3
Histogram	10	8
Texture	28	8
Moments	14	7
Radon	20	13
Granulometric	12	9
Shape, Histogram	13	10
Shape, Texture	31	9
Shape, Moments	17	8
Shape, Radon	23	14
Shape, Granulometric	15	11
Shape, Histogram, Texture	41	13
Shape, Histogram, Moments	27	14
Shape, Histogram, Radon	33	16
Shape, Histogram, Granulometric	25	15
Shape, Texture, Moments	45	12
Shape, Texture, Radon	51	16
Shape, Texture, Granulometric	43	13
Shape, Moments, Radon	37	15
Shape, Moments, Granulometric	29	13
Shape, Radon, Granulometric	35	17
Shape, Histogram, Texture, Moments	55	15
Shape, Histogram, Texture, Radon	61	17
Shape, Histogram, Texture, Granulometric	53	16
Shape, Histogram, Moments, Radon	47	16
Shape, Histogram, Moments, Granulometric	39	17
Shape, Histogram, Radon, Granulometric	45	18
Shape, Texture, Moments, Radon	65	17
Shape, Texture, Moments, Granulometric	57	14
Shape, Texture, Radon, Granulometric	63	17
Shape, Moments, Radon, Granulometric	49	19
Shape, Histogram, Texture, Moments, Radon	75	18
Shape, Histogram, Texture, Moments, Granulometric	67	17
Shape, Histogram, Moments, Radon, Granulometric	59	18
Shape, Histogram, Texture, Radon, Granulometric	73	18
Shape, Texture, Moments, Radon, Granulometric	77	18
Shape, Histogram, Texture, Moments, Radon, Granulometric	87	18

CHAPTER 5

Content-based Retrieval of Mammograms

5.1 QUERY AND COMPARISON OF IMAGES FOR CBIR

Based on the methods described in the preceding chapters, we can now represent various character-istics of mammograms with a set of features and index the images with a reduced set of parameters obtained via PCA. In this manner, the large number of pixels in a mammogram are reduced to a small number of parameters. The set of parameters may be treated as a vector that characterizes the mammogram for further analysis. The vector serves as an index that represents the corresponding image for CBIR or CAD. In the present chapter, we shall study methods to compare and retrieve images according to certain specifications.

Figure 5.1 shows a schematic representation of the CBIR system developed in the present work [11, 12]. In order to retrieve images from a database in response to a query, a comparison is performed between the feature vector of the query image and the corresponding vectors of all of the images in the database. The comparison is made using measures of similarity in the space of metrics related to the features. Several functions exist for use as measures of similarity of vectors. The most

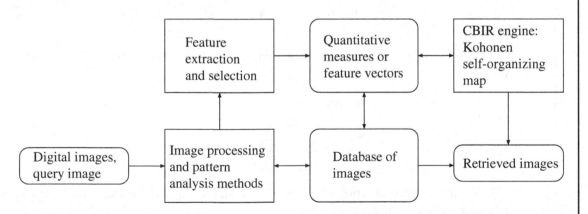

Figure 5.1: Schematic representation of a CBIR system. The search engine used is based on a neural network using a Kohonen self-organizing map. Based on a similar figure in Kinoshita et al. [11].

commonly used measure is the Euclidean distance. Müller et al. [8] provide a detailed review of CBIR.

A query or search of similarity could be made in two ways: range query and k-nearest-neighbor (k-NN) query. A range query uses a limit for the distance from the query sample or a reference point to retrieve images: all images in the database having a distance less than the specified limit are retrieved and presented to the user. In a k-NN query, the k-nearest images to the query image are retrieved. In a practical medical application, the user may be interested in only a small number of images, usually an odd number, such as 3, 5, or 7, for comparative analysis with the current image on hand (the query case).

A CBIR system possesses a structure similar to that of a CAD system, except for the difference in the final objective. A CBIR system uses quantitative features for indexing, comparative analysis, and retrieval of images based upon a similarity measure. A CAD system may use similar measures and methods, but with the aim of leading toward a diagnostic decision. Regardless, both CBIR and CAD systems may be used as diagnostic aids.

5.1.1   MEASURES OF DISTANCE AND SIMILARITY

Consider the representation of an image using a vector of features, $\mathbf{f}$, where each vector has Q elements or parameters, that is, $\mathbf{f} = [f_1, f_2, \ldots, f_Q]^T$. Let the feature vector of the query image being used as the reference to perform CBIR be $\mathbf{q} = [q_1, q_2, \ldots, q_Q]^T$. To compare the query image with an image from the database using their feature vectors, we need a measure of vectorial distance. A few examples of such measures are described below [3, 116]:

- Euclidean distance

$$D_E^2 = \|\mathbf{f} - \mathbf{q}\|^2 = (\mathbf{f} - \mathbf{q})^T (\mathbf{f} - \mathbf{q}) = \sum_{i=1}^{Q} (f_i - q_i)^2. \tag{5.1}$$

A small value of D_E indicates greater similarity between the two images or vectors than a large value of D_E.

- Manhattan or city-block distance

$$D_C = \sum_{i=1}^{Q} |f_i - q_i|. \tag{5.2}$$

The Manhattan distance is the shortest path between $\mathbf{f}$ and $\mathbf{q}$, with each segment being parallel to a coordinate axis [114].

- Normalized dot product

$$D_d = \frac{\mathbf{f}^T \mathbf{q}}{\|\mathbf{f}\| \|\mathbf{q}\|}. \tag{5.3}$$

A large dot product value indicates a greater degree of similarity between the two vectors than a small value. The measure may be reversed to serve as distance by considering $(1 - D_d)$.

A distance measure may be easily converted to a measure of similarity: if a distance measure D is normalized to the range $[0, 1]$, a measure of similarity is given by $(1 - D)$.

5.1.2   THE KOHONEN SELF-ORGANIZING MAP

In the self-organizing map (SOM) architecture of neural networks, neurons are organized in the form of rectangular or hexagonal grids, and are usually arranged in 2D layers. Figure 5.2 presents the topology of a SOM network and the grids showing the neighborhood of a "winner neuron." In the basic scheme of the SOM network, the neurons of the output layer compete for the information presented at the neurons of the input layer [117]. If a winner neuron emerges, that neuron is readjusted to respond in a stronger manner to the same stimulus. Within this unsupervised model, not only does the winner neuron get adjusted, its neighbors are also adjusted.

A SOM network functions as follows. The synaptic weights are initialized with small random values. An input signal $\mathbf{x} = [x_1, x_2, \ldots, x_N]$, with its values representing an item of information, is provided to the network, without any desired output being specified (that is, unsupervised training). Given an input $\mathbf{x}$, an output neuron j will provide the highest response; this neuron is considered to be the winner. This neuron is always activated with a high response whenever the same input pattern is applied to the neurons in the input layer. The winner neuron j and its neighbors $\mathbf{V}_j$ have their synaptic weights adjusted so as to provide a stronger response to the input $\mathbf{x}$. The network is considered to be trained after all of the training data have been presented to the network, and the training criteria have been satisfied.

After the training phase, the network may be used for query and retrieval purposes. This phase is essentially the same as the training phase, except that the weights of the neurons are not altered any more. When the test entries are recognized correctly, the network is considered to have been successfully trained.

The adaptation of the neurons is crucial for orderly training of the SOM, because, despite the fact that each neuron is modified independently of the others, it is the whole set of neurons that represents the information conveyed by the input. The adjustment of the synaptic weights takes place in the following manner. Initially, the winner neuron is found for the input presented; this is the neuron having the least distance between its weight vector $\mathbf{w}_j$ and the input vector $\mathbf{x}$. In each step of the training phase, the neuron with the highest response is adjusted so as to provide an even larger response for the same input; simultaneously, the neurons in the neighborhood $\mathbf{V}_j$ are also adjusted. The neurons that are outside the subset of the neighborhood as defined above are not adjusted.

The adaptation of the neurons is performed as follows: once the winner neuron is selected, the weights of the neuron and its neighbors are updated as

$$\mathbf{w}_j(t+1) = \begin{cases} \mathbf{w}_j(t) + \alpha(t)|\mathbf{x} - \mathbf{w}_j(t)| & \text{if } j \in \mathbf{V}_j(t) \\ \mathbf{w}_j(t) & \text{if } j \notin \mathbf{V}_j(t) \end{cases}, \tag{5.4}$$

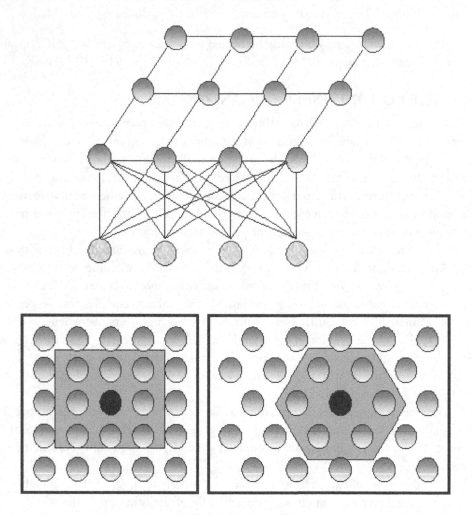

Figure 5.2: Architecture of the Kohonen SOM network and neighborhoods in rectangular and hexagonal grids. The winner neurons are shown in black and the neighbors are shown in gray. Neighborhoods of radius $R = 0, 1$, and 2 are shown for rectangular and hexagonal grids in the lower part of the figure. Reproduced with kind permission from Springer Science+Business Media B. V. from S. K. Kinoshita, P. M. de Azevedo-Marques, R. R. Pereira Jr., J. A. H. Rodrigues, and R. M. Rangayyan, "Content-based retrieval of mammograms using visual features related to breast density patterns," Journal of Digital Imaging, 20(2): 172–190, June 2007. © Springer.

where t is the stage of iteration, $\mathbf{V}_j$ is the neighborhood of neurons, and α is the rate of learning.

Kohonen [117] proposed that, in a practical application, the training process could be stopped when there is no further change in the weights of the selected neuron. The process may also be stopped after a certain number of iterations (epochs).

Laaksonen et al. [118] described the use of the SOM as a CBIR system for retrieval of multimodality images. In their work, a SOM network is trained in the unsupervised mode with a set of features. The similarity between images depends upon the characteristics of the input features and the measure of similarity used. Then similar images are grouped together in classes according to a threshold applied to the selected measure of similarity. For every image provided as the query case, the features of the image point toward one of the classes of images. If the image cannot be grouped with any of the classes, it is used to form a new class.

The structure of the Kohonen SOM network used in the experiments conducted in the present work varied as follows [11]. The number of input neurons was set to be equal to the number of features used to represent a mammogram, as listed in Table 4.3 for the various feature combinations evaluated. A rectangular grid was used for creating a 17×17-neighborhood matrix of neurons for the output layer (289 neurons). The distance D between two neurons was defined as the magnitude of the maximum element in the vector obtained by subtracting the neurons' weight vectors. Selection of neighboring neurons was made possible with a radius varying from 0 to 16 around the winning neuron. The neighborhoods of radius $R = 0, 1$, and 2 are shown in Figure 5.2. A maximum of 8,000 cycles was used in the training process. The learning rate was not fixed, but did not exceed 0.0001 in any of the tests.

The leave-one-out method was used to create the training and testing cases from the available database of mammograms. The database used was manually divided into four groups of 270 images each (left CC, right CC, left MLO, and right MLO). In this procedure, for each group, a query image is separated from the data set containing 270 images and the SOM network is trained with the 269 remaining images. Then the query image is used as the system input. This procedure is repeated until all images are used as the query. Because of the use of an output layer layout with 289 neurons, every image can be classified as a single class.

To evaluate the performance of the CBIR system, measures of precision and recall were obtained according to the rules given in Table 5.1 for comparing the BI-RADS categorization (performed by the radiologists) between the query and the retrieved mammograms, as follows:

$$\text{precision} = \frac{\text{NRR}}{\text{TR}} \tag{5.5}$$

and

$$\text{recall} = \frac{\text{NRR}}{\text{TRC}}, \tag{5.6}$$

where NRR is the number of relevant images retrieved, TR is the total number of images retrieved, and TRC is the total number of relevant images in the database. For each CBIR experiment, a range query was applied for image retrieval using SOM radius distance from $R = 0$ to $R = 16$, and the average and standard deviation values of precision were computed.

Table 5.1: Rules used for verification of the precision of retrieval, taking into consideration the potential variability in evaluation of breast tissue composition using BI-RADS categorization. Reproduced with kind permission from Springer Science+Business Media B. V. from S. K. Kinoshita, P. M. de Azevedo-Marques, R. R. Pereira Jr., J. A. H. Rodrigues, and R. M. Rangayyan, "Content-based retrieval of mammograms using visual features related to breast density patterns," Journal of Digital Imaging, 20(2): 172–190, June 2007. © Springer.

BI-RADS category of query image	Acceptable categories of retrieved images
1	1 and 2
2	1, 2, and 3
3	2, 3, and 4
4	3 and 4

5.1.3   RESULTS OF CBIR WITH MAMMOGRAMS

Table 5.2 lists the precision (mean and standard deviation) obtained for different combinations of features considering the whole set of images, precision considering that 25% of images are retrieved, and precision when 50% of images are retrieved. Various combinations of features provided average precision rates in the range from 79% to 83%. The results obtained indicate that the proposed features can facilitate CBIR of mammograms with good levels of accuracy. However, the nature of similarity represented by the various features and their combinations differ substantially, and an appropriate set of features needs to be selected based upon the desired results. Good feature selection can facilitate the clustering of images into the different categories of interest.

Figure 5.3 shows an example of a query image and retrieved images based upon the combination of shape, histogram, and moment-based features. Figure 5.4 shows an example using the combination of shape, histogram, texture, and granulometric features. Figure 5.5 shows an example using the combination of shape, histogram, moment-based, Radon-domain, and granulometric features. The feature combinations mentioned above provided average precision values of 82.27%, 81.32%, and 82.45%, respectively. It can be observed that the retrieved images match in size and shape with the query images, which was expected because of the effectiveness of shape features for CBIR.

The SOM average training time for each leave-one-out cycle was 62.6 s. Testing time (retrieval time) was directly related with the range query applied, varying from 0.1 ms for radius equal to 0 to 562.3 ms for radius equal to 16.

Table 5.2: Mean and standard deviation of precision, the precision with the first 25% of the images retrieved, and precision with the first 50% of the images retrieved, obtained for the various groups and selected combinations of features. Texture: Haralick's texture features. Histogram: Statistics of gray-level histogram. Moments: Hu's moment-based features. Radon: Radon-domain features. Shape: Shape features. Granulometric: Granulometric features. Reproduced with kind permission from Springer Science+Business Media B. V. from S. K. Kinoshita, P. M. de Azevedo-Marques, R. R. Pereira Jr., J. A. H. Rodrigues, and R. M. Rangayyan, "Content-based retrieval of mammograms using visual features related to breast density patterns," Journal of Digital Imaging, 20(2): 172–190, June 2007. © Springer.

Features	Precision (%)	At 25%	At 50%
Shape	79.15 ± 1.11	80.55	79.93
Histogram	81.20 ± 3.02	82.77	80.45
Texture	81.01 ± 2.64	82.33	80.68
Moments	79.16 ± 1.30	79.92	79.00
Radon	79.35 ± 1.82	79.66	78.66
Granulometric	79.56 ± 1.95	80.75	78.93
Shape, Histogram	82.28 ± 3.76	84.17	82.10
Shape, Texture	80.86 ± 2.78	82.33	80.55
Shape, Moments	80.15 ± 1.58	81.33	80.48
Shape, Radon	80.45 ± 2.68	81.29	80.04
Shape, Granulometric	79.43 ± 1.99	79.86	78.96
Shape, Histogram, Texture	81.77 ± 3.76	83.84	81.37
Shape, Histogram, Moments	82.27 ± 4.11	85.55	82.10
Shape, Histogram, Radon	81.29 ± 3.21	83.10	81.07
Shape, Histogram, Granulometric	79.96 ± 2.95	80.98	78.71
Shape, Texture, Moments	80.66 ± 2.32	82.35	80.39
Shape, Texture, Radon	80.84 ± 2.96	82.50	80.56
Shape, Texture, Granulometric	80.14 ± 2.49	81.43	79.26
Shape, Moments, Radon	80.26 ± 2.45	81.08	79.59
Shape, Moments, Granulometric	79.38 ± 1.68	79.90	78.96
Shape, Radon, Granulometric	80.10 ± 2.11	81.26	79.48
Shape, Histogram, Texture, Moments	81.97 ± 3.94	84.31	81.48
Shape, Histogram, Texture, Radon	82.67 ± 4.04	85.22	82.93
Shape, Histogram, Texture, Granulometric	81.32 ± 3.80	83.01	80.31
Shape, Histogram, Moments, Radon	82.06 ± 3.60	84.52	81.69
Shape, Histogram, Moments, Granulometric	80.50 ± 2.59	81.61	79.72
Shape, Histogram, Radon, Granulometric	80.78 ± 3.42	82.65	80.23
Shape, Texture, Moments, Radon	81.72 ± 3.28	83.90	81.45
Shape, Texture, Moments, Granulometric	81.00 ± 2.97	82.27	80.68
Shape, Texture, Radon, Granulometric	81.14 ± 3.25	82.65	80.29
Shape, Moments, Radon, Granulometric	80.01 ± 1.99	81.17	79.63
Shape, Histogram, Texture, Moments, Radon	82.44 ± 3.99	85.41	82.83
Shape, Histogram, Texture, Moments, Granulometric	81.60 ± 3.70	83.96	80.76
Shape, Histogram, Moments, Radon, Granulometric	82.45 ± 3.46	84.67	82.35
Shape, Histogram, Texture, Radon, Granulometric	81.04 ± 3.14	82.14	80.50
Shape, Texture, Moments, Radon, Granulometric	82.14 ± 3.67	84.29	81.10
Shape, Histogram, Texture, Moments, Radon, Granulometric	81.62 ± 3.92	84.03	80.32

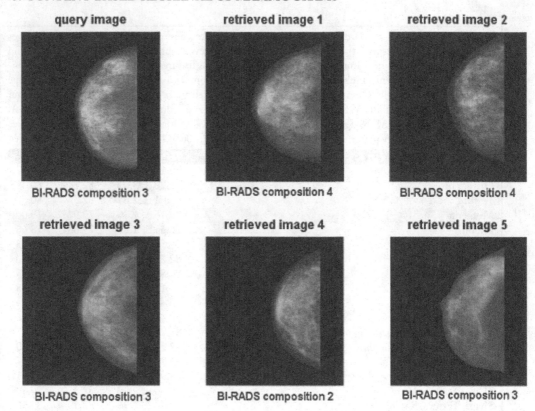

Figure 5.3: Example of CBIR using the combination of shape, histogram, and moment-based features. Range query with radius $R = 0$.

An important characteristic of the methodology implemented in this work is the use of an a priori classification based on breast density composition performed by radiologists as an initial stage for query training for image similarity. In an a posteriori stage, the system could be improved using RFb based on visual inspection by radiologists to determine the degree of similarity between the query and retrieved images (see Section 5.2). A supervised neural network could be trained to match the perception of similarity by radiologists; this could lead to improved performance, as suggested by Muramatsu et al. [16].

Figures 5.6 to 5.10 show precision-versus-recall curves for various combinations of groups of features. Although the initial values of precision are high in each case, it is evident that precision decreases as recall increases. However, it should be noted that, in a medical imaging application, it is not necessary to retrieve all of the cases in the database that are similar to the query case. It is adequate

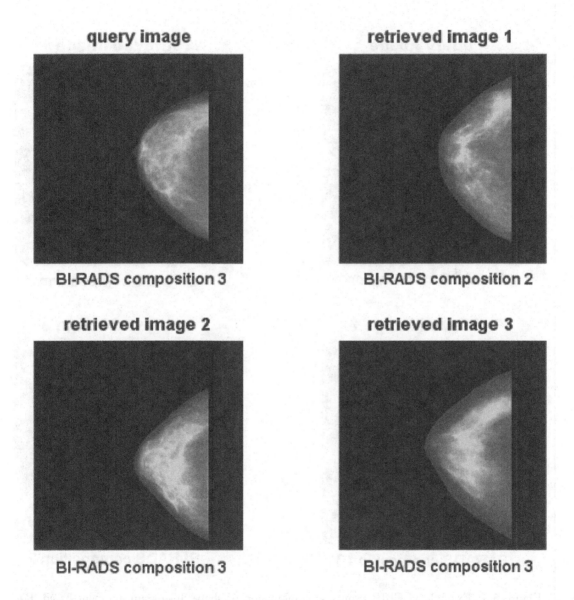

Figure 5.4: Example of CBIR using the combination of shape, histogram, texture, and granulometric features. Range query with radius $R = 0$.

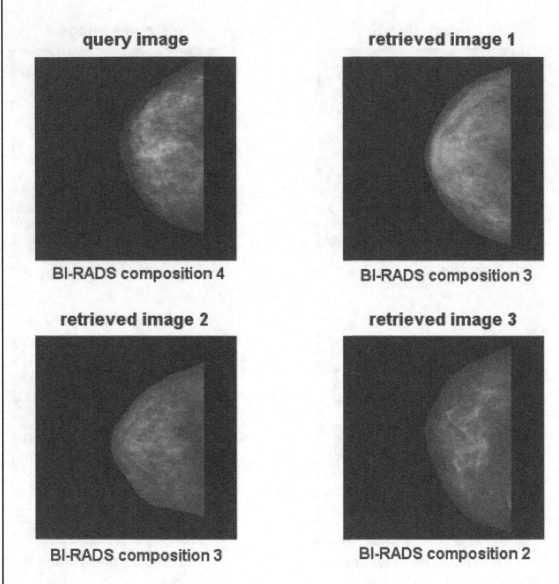

Figure 5.5: Example of CBIR using the combination of shape, histogram, moment-based, Radon-domain, and granulometric features. Range query with radius $R = 0$.

if a small number, preferably an odd number, such as 3, 5, or 7, of images that most closely match the query are retrieved. The radiologist or physician could then review the small number of retrieved cases. A radiologist would not want to review all of the cases in the database that are similar to the query case. In consideration of this preferred mode of operation, the precision-versus-recall curve that is commonly used to evaluate general CBIR systems may not be suitable for the assessment of CBIR systems for medical applications. Regardless, the high precision achieved by the proposed methods is evident at the beginning of each precision-versus-recall curve illustrated.

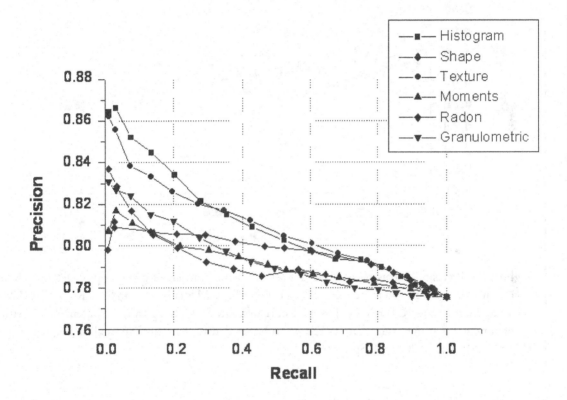

Figure 5.6: Precision-versus-recall curves for individual groups of features. Reproduced with kind permission from Springer Science+Business Media B. V. from S. K. Kinoshita, P. M. de Azevedo-Marques, R. R. Pereira Jr., J. A. H. Rodrigues, and R. M. Rangayyan, "Content-based retrieval of mammograms using visual features related to breast density patterns," Journal of Digital Imaging, 20(2): 172–190, June 2007. © Springer.

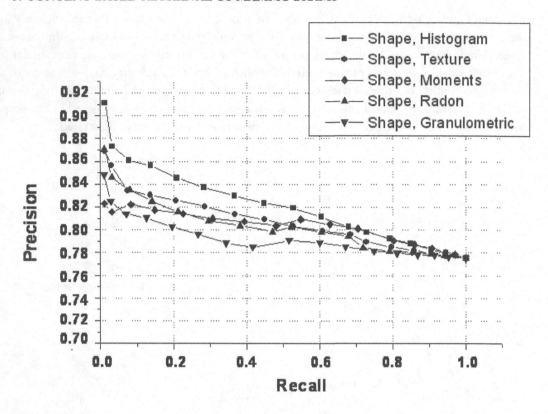

Figure 5.7: Precision-versus-recall curves for selected combinations of two groups of features. Reproduced with kind permission from Springer Science+Business Media B. V. from S. K. Kinoshita, P. M. de Azevedo-Marques, R. R. Pereira Jr., J. A. H. Rodrigues, and R. M. Rangayyan, "Content-based retrieval of mammograms using visual features related to breast density patterns," Journal of Digital Imaging, 20(2): 172–190, June 2007. © Springer.

5.2   CBIR WITH RELEVANCE FEEDBACK

Azevedo-Marques et al. [12] implemented a CBIR system with RFb, which is described in the present section. The CBIR system was developed using Borland C++BuilderVersion 5.0 and Oracle 9.1, including techniques for RFb, with access to the RIS of the Clinical Hospital of the Faculty of Medicine of Ribeirão Preto. The CBIR system facilitates the selection of a query image, the type of query (kNN or range) and the related parameters, and the feature vector. The results of retrieval were presented to the user in the format displayed in Figure 5.11.

In the CBIR system, to facilitate RFb, the user is provided with a window to label each retrieved image as highly relevant, relevant, or not relevant, with the associated weights of relevance

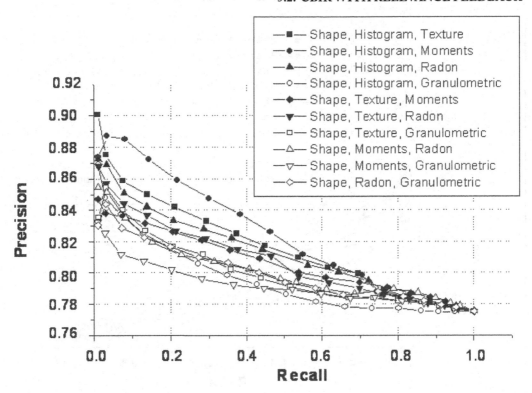

Figure 5.8: Precision-versus-recall curves for selected combinations of three groups of features. Reproduced with kind permission from Springer Science+Business Media B. V. from S. K. Kinoshita, P. M. de Azevedo-Marques, R. R. Pereira Jr., J. A. H. Rodrigues, and R. M. Rangayyan, "Content-based retrieval of mammograms using visual features related to breast density patterns," Journal of Digital Imaging, 20(2): 172–190, June 2007. © Springer.

of 2, 1, and 0, respectively. The criteria used for labeling the retrieved images are listed in Table 5.3. In the experiments conducted, the radiologists did not refer to the RIS reports; the images were interpreted independently, in terms of the distribution of density and composition of the breast. When the retrieval process is iterated, the retrieved images that have already been labeled with a weight in a preceding iteration are displayed with their weights to obviate repeated assignment of the weight of relevance; images labeled previously as not relevant are not retrieved again.

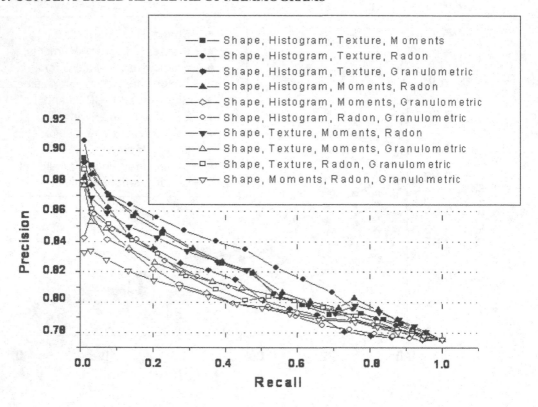

Figure 5.9: Precision-versus-recall curves for selected combinations of four groups of features. Reproduced with kind permission from Springer Science+Business Media B. V. from S. K. Kinoshita, P. M. de Azevedo-Marques, R. R. Pereira Jr., J. A. H. Rodrigues, and R. M. Rangayyan, "Content-based retrieval of mammograms using visual features related to breast density patterns," Journal of Digital Imaging, 20(2): 172–190, June 2007. © Springer.

5.2.1 METHODS OF RELEVANCE FEEDBACK

The CBIR system described in the work of Azevedo-Marques et al. [12] provides options for the method of RFb, including the method of query point movement (QPM), relevance feedback projection (RFP), and relevance feedback using multiple point projection (MPP) [119, 120]. QPM, as proposed by Rocchio [119], changes the query center to a new one, which is nearer to the images selected as relevant (positive objects) and farther from the images marked as irrelevant (negative objects). In this manner, it becomes possible to amend the query and incorporate the user's judgment.

RFP and MPP are modifications of the original QPM method that was proposed by Traina et al. [120]. RFP is based on the projection of the range given by the irrelevant images on to a relevant

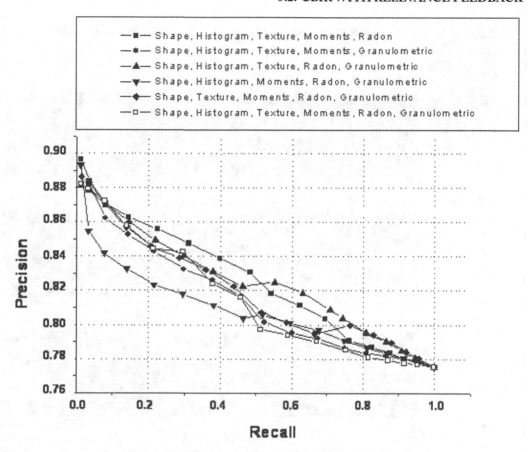

Figure 5.10: Precision-versus-recall curves for selected combinations of five and six groups of features. Reproduced with kind permission from Springer Science+Business Media B. V. from S. K. Kinoshita, P. M. de Azevedo-Marques, R. R. Pereira Jr., J. A. H. Rodrigues, and R. M. Rangayyan, "Content-based retrieval of mammograms using visual features related to breast density patterns," Journal of Digital Imaging, 20(2): 172–190, June 2007. © Springer.

interval. RFP analyzes each feature (image attribute) from the feature vector separately, and classifies it according to a relevance rule, taking into account the relevant and irrelevant images. The values of the attribute being processed are placed in a relevance interval that is used as the reference for the generation of a new weighted feature vector. A new query is formed by using the new feature vector as the query center. In this way, information obtained from the images selected as irrelevant is applied to move the query center away from the negative objects.

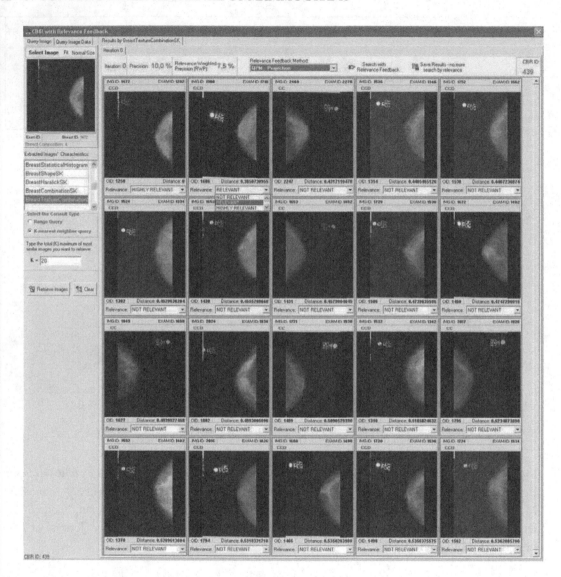

Figure 5.11: Illustration of the display of results of CBIR with a query image, including the assignment of relevance of the retrieved images. Reproduced with kind permission from Springer Science+Business Media B. V. from P. M. de Azevedo-Marques, N. A. Rosa, A. J. M. Traina, C. Traina Junior, S. K. Kinoshita, and R. M. Rangayyan, "Reducing the semantic gap in content-based image retrieval in mammography with relevance feedback and inclusion of expert knowledge," International Journal of Computer Assisted Radiology and Surgery, 3(1-2): 123–130, June 2008. © Springer.

Table 5.3: Criteria for automatic assessment of the relevance of the retrieved images in accordance with the BI-RADS categories and the associated weights. The assessment of the relevance of the retrieved images by using this table, in comparison with the BI-RADS categorization in the official RIS report, is referred to as "RADR" in the text. Reproduced with kind permission from Springer Science+Business Media B. V. from P. M. de Azevedo-Marques, N. A. Rosa, A. J. M. Traina, C. Traina Junior, S. K. Kinoshita, and R. M. Rangayyan, "Reducing the semantic gap in content-based image retrieval in mammography with relevance feedback and inclusion of expert knowledge," International Journal of Computer Assisted Radiology and Surgery, 3(1-2): 123–130, June 2008. © Springer.

Query	Retrieved BI-RADS 1	BI-RADS 2	BI-RADS 3	BI-RADS 4
BI-RADS 1	Highly relevant (weight=2)	Relevant (weight=1)	Not relevant (weight=0)	Not relevant (weight=0)
BI-RADS 2	Relevant (weight=1)	Highly relevant (weight=2)	Relevant (weight=1)	Not relevant (weight=0)
BI-RADS 3	Not relevant (weight=0)	Relevant (weight=1)	Highly relevant (weight=2)	Relevant (weight=1)
BI-RADS 4	Not relevant (weight=0)	Not relevant (weight=0)	Relevant (weight=1)	Highly relevant (weight=2)

MPP processes a query centered at each image selected as relevant. An additional image (phantom) is generated employing the new weighted feature vector. A query is also performed by considering the phantom as its center. A new combined distance is computed based on the distance between the resulting object set as the corresponding query center, considering each initial relevant image, and the phantom. The images with the smaller combined distance are returned as the result of the query. All of the information regarding the queries, the parameters used, and the images retrieved (along with their distance measures with respect to the query image and the weight of relevance provided by the user) are saved for further analysis.

5.2.2   RELEVANCE-WEIGHTED PRECISION OF RETRIEVAL

In the work of Azevedo-Marques et al. [12], the precision of retrieval was computed as the ratio of the number of relevant images retrieved to the total number of images retrieved. In computing precision, each retrieved image received a binary weight of unity or zero, representing a relevant image or not; the sum of such scores for all of the retrieved images was divided by the total number of retrieved images. In addition, a measure of relevance-weighted precision (RWP) was proposed, computed by applying the weights as listed in Table 5.3 to the retrieved images. The sum of the scores for all of the retrieved images was normalized by twice the total number of images retrieved; the latter value represents the best retrieval situation where all of the retrieved images receive the maximal weight of two.

5.2.3   ASSESSMENT OF THE BENEFITS OF RFB IN CBIR

Three radiologists, referred to as RAD1, RAD2, and RAD3, conducted CBIR experiments with the system developed by Azevedo-Marques et al. [12]. The kNN search was employed, with $k = 20$, using one mammographic image of each BI-RADS breast density index as the query image. The query images were selected by one of the radiologists; the same query images were used in the experiments conducted by the other two radiologists.

In addition to the above, a CBIR experiment was conducted with the same set of query images as mentioned above, with automatic classification of the relevance of the results in accordance with the BI-RADS breast density index entered in the official RIS report (see Table 5.3). Although the RIS reports had been entered by several different radiologists at various points in time, they were considered as a single expert radiologist, referred to as RADR, in the analysis of CBIR and RFb.

The experiments mentioned above were iterated, with the query point being moved at the end of each iteration, using the RFP method described in Section 5.2.1, to include RFb for the subsequent iteration. The results of CBIR were displayed in a new screen after each iteration, for analysis by the user.

Figure 5.11 shows a screen with the display of the results of CBIR. The query image and 20 retrieved images are shown in the figure. The process of application of the weight of relevance by the user is also illustrated. Figure 5.12 shows a screen with the display of the results of CBIR shown in Figure 5.11 after the RFb procedure. It is possible to see the improvement of precision and RWP.

Figure 5.13 shows the precision and RWP of retrieval obtained for the various experiments conducted; the first trial of CBIR before RFb is termed as "Iteration 0." The mean and standard deviation of the precision and RWP for five iterations with RFb are shown in Figure 5.13. Figure 5.14 shows precision and RWP gain/loss for each category of images used in the test.

With the query image of BI-RADS breast density index 1, there was no gain due to RFb; instead, there was a small amount of loss. The precision of retrieval was nearly 100% without RFb; this could be due to the ease of representation and analysis of images with low density of the breast. The gain in precision or RWP due to RFb increased with increase in overall breast density, as indicated by the BI-RADS breast density indices of 2, 3, and 4. It is evident from Figure 5.13 that, while the precision of retrieval in Iteration 0, on the average, is the lowest in the case of the query image being of BI-RADS breast density index 4, the gain in precision or RWP is the highest; this indicates that RFb can serve better in CBIR and analysis of difficult cases related to dense breasts.

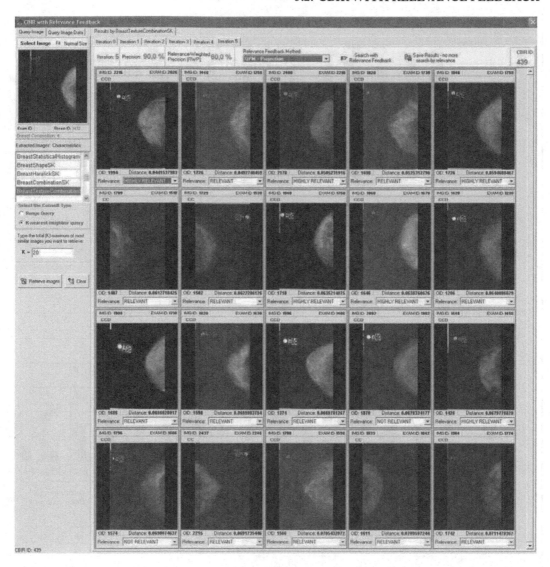

Figure 5.12: Illustration of the display of results of CBIR of the same retrieval presented in Figure 5.11 after the RFb procedure. The improvement in the precision and RWP is evident. Reproduced with kind permission from Springer Science+Business Media B. V. from P. M. de Azevedo-Marques, N. A. Rosa, A. J. M. Traina, C. Traina Junior, S. K. Kinoshita, and R. M. Rangayyan, "Reducing the semantic gap in content-based image retrieval in mammography with relevance feedback and inclusion of expert knowledge," International Journal of Computer Assisted Radiology and Surgery, 3(1-2): 123–130, June 2008. © Springer.

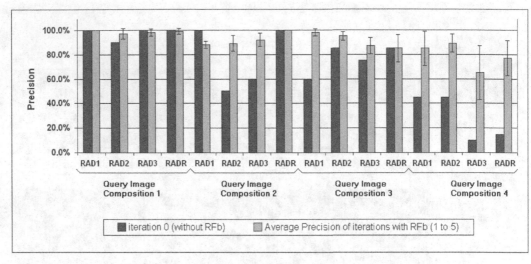

(a)

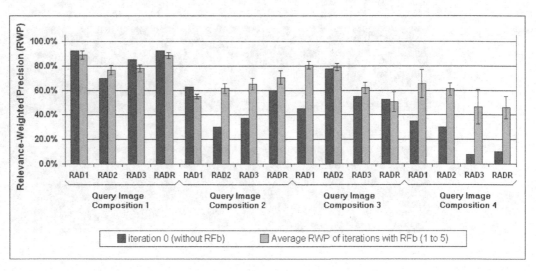

(b)

Figure 5.13: Plots of precision and RWP of retrieval for some of the CBIR experiments conducted. Reproduced with kind permission from Springer Science+Business Media B. V. from P. M. de Azevedo-Marques, N. A. Rosa, A. J. M. Traina, C. Traina Junior, S. K. Kinoshita, and R. M. Rangayyan, "Reducing the semantic gap in content-based image retrieval in mammography with relevance feedback and inclusion of expert knowledge," International Journal of Computer Assisted Radiology and Surgery, 3(1-2): 123–130, June 2008. © Springer.

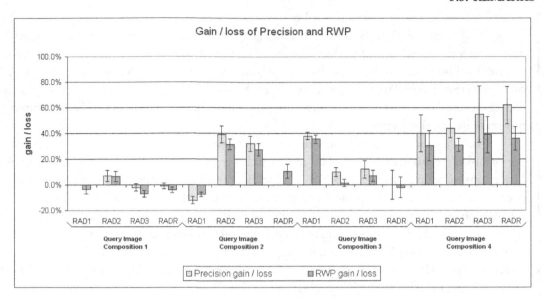

Figure 5.14: Plots of precision and RWP gain or loss for each BI-RADS query image composition. Reproduced with kind permission from Springer Science+Business Media B. V. from P. M. de Azevedo-Marques, N. A. Rosa, A. J. M. Traina, C. Traina Junior, S. K. Kinoshita, and R. M. Rangayyan, "Reducing the semantic gap in content-based image retrieval in mammography with relevance feedback and inclusion of expert knowledge," International Journal of Computer Assisted Radiology and Surgery, 3(1-2): 123–130, June 2008. © Springer.

5.3   REMARKS

Regardless of the fact that the key features that need to be assessed in diagnosing breast pathology are microcalcifications and masses, the evaluation of breast density (high-density fibroglandular tissue) and asymmetry plays an important role in the diagnostic process. In fact, the evaluation of bilateral asymmetry based on density, shape, and size is usually the first stage in the mammographic evaluation process. Radiologists look for calcifications, masses, asymmetry, and architectural distortion in mammograms: a CBIR system for mammography should include the possibility of query by similarity based on all of these features [9, 10]. A hierarchical structure should be implemented for the characterization of breast images beginning with general structural features, such as size, shape, and density, and continuing with more specific features associated with radiological findings such as contours of masses and the number as well as spatial distribution of microcalcifications in a cluster. Thus, a query for CBIR may be performed based on the similarity between general structural features of the breast, similarity between features associated with a specific radiological finding, or similarity between ROIs in mammograms.

The concept of similarity is notably important because no two mammograms or ROIs may be expected to be identical, even when belonging to the same diagnostic category, and a perfect or exact match to a query is improbable. The use of vectors of quantitative parameters to index image features facilitates the use of simple distance measures (such as the Euclidean distance) to select the cases that are most similar to the query sample. In this context, a study of several types of quantitative features focused on breast density, size, texture, and shape for application in CBIR of mammograms was presented in this chapter.

The evaluation of the performance of the proposed procedures for CBIR was based upon the precision-versus-recall curve. The results indicate that, while the statistics of the gray-level histograms of subregions of the breast in mammograms can provide high levels of precision for mammographic image retrieval based on structural features, the results can be improved further by the inclusion of shape, Radon-domain, moment-based, texture, and granulometric features.

Although there exist more sophisticated methods for characterization of breast density, such as fractal analysis [121–123], the methods applied in the present work have provided good results. A semantic gap will be present even when using more sophisticated criteria for image retrieval based on breast density; the RFb methodology presented in this chapter should be able to reduce the semantic gap in such situations as well. In this context, the demonstration and assessment of the potential of RFb to reduce the semantic gap are the main topics of concern of the present chapter.

It is important to note the approach used for the evaluation of the performance of the proposed procedures for CBIR, which is based upon the precision rate. Receiver operating characteristic (ROC) curves are suitable for the evaluation of the performance of a CAD system that involves a diagnostic decision, in terms of sensitivity and specificity. However, for the evaluation of the performance of a CBIR system that involves information retrieval, it is more appropriate to use precision as a quantitative parameter. In an information retrieval task, the measure of precision is an important indicator of the similarity of the retrieved images with respect to the query sample. In the context of CBIR, precision gives us a measure of the user's satisfaction with the retrieved images. In a clinical scenario in mammography, a radiologist can use a CBIR system to assist in the classification of an image based on the BI-RADS breast density index. If the query results bring back images with BI-RADS breast density index 3 and 4, for example, the radiologist will be able to analyze them, determine which images are similar to the query image, evaluate the classification of the retrieved images, and decide on the classification of the image on hand. The radiologist could also gain an understanding on why some of the retrieved images have been classified with density index 3 and others with density index 4. In this manner, the CBIR system can assist in arriving at a classification decision when the image on hand is at the boundaries between different classes.

Furthermore, the issue of RFb in the present context is not related to the consistency of the quantitative results provided by the feature extraction methods, but related to the question of how the quantitative results can be associated with the expert user's visual perception. Therefore, in the study illustrated in the present chapter, the assessments provided by the radiologists (experts in mammography) were used as the gold standard, despite the potential for interobserver and intraobserver variability.

Although the results presented in this work are related to breast density classification in mammographic images, other feature extractors designed to characterize masses [10] and microcalcification clusters [9] can be incorporated into the CBIR system. The RFb methodology described in the present work is generic, and can be applied to other imaging modalities and clinical applications, provided an appropriate set of feature extractors is available.

We have presented a CBIR system with RFb for application in the analysis of mammograms. The system includes several features related to texture, distribution of breast density, and shape to index the images in the database, as well as techniques to incorporate the indication of relevance of the retrieved images provided by a user. The results indicate that RFb can improve the precision of retrieval. The proposed methods could find applications in CAD of breast cancer.

The next chapter presents methods and models to facilitate the incorporation of CBIR and CAD into PACS, RIS, and HIS in a practical clinical environment.

CHAPTER 6

Integration of CBIR and CAD into Radiological Workflow

6.1   THE HOSPITAL INFORMATION SYSTEM

The HIS can be understood as a management system that, on the one hand, is able to support patient-care activities in the hospital and, on the other hand, provides administrative functionalities to support the hospital's daily business transactions. Because of the nature of the information that is handled by the HIS, it is also an important tool for evaluation of the costs and performance of the hospital, and can be used for long-term planning. Commonly, the HIS is an umbrella for other more specific information systems, such as the RIS for radiology and others for pathology, pharmacy, and clinical laboratories. In this sense, the HIS propagates, in real time, patient demographic information to several other specific systems. Also, the HIS must provide automation for such events as patient registration, admissions and transfer, patient accounting, and on-line access to clinical results [124].

6.2   THE RADIOLOGY INFORMATION SYSTEM

The RIS must support both the administrative and clinical operation of a radiology department. The main objectives of the RIS are to reduce administrative overhead and improve the quality of delivery of radiological examinations [124]. Computer-based RIS has been developed to handle almost the entire spectrum of information-management tasks in a radiology department, such as scheduling of examinations; registration of patients; performance of examinations; interpretation, review, and analysis of studies by radiologists; distribution of radiology reports; billing for services; and maintenance of archives of digital exams [125]. The configuration of a RIS is similar to that of a HIS, except that it is on a smaller scale [124].

6.3   PICTURE ARCHIVAL AND COMMUNICATION SYSTEMS

Information systems for managing medical images and data first appeared in the late 1980s, when digital acquisition procedures began to be used in large hospitals. At that time, each procedure was considered as an isolated system, simply connected to a local display station and a dedicated printer [126, 127]. The potential of such early systems, along with the growth of procedures for distribution of digital information, produced the need to standardize the workflow and the format for storage and communication of medical images [126]. In response to this need, the National Electrical Manufacturers Association (NEMA) and the ACR, with the support of the Radiological Society of North America (RSNA), a number of universities, and several manufacturers of radiological equipment, created the concept of PACS [128].

Typically, a PACS is a combination of hardware and software dedicated to short- and long-term storage, retrieval, management, distribution, and presentation of images. A PACS can be divided into four subsystems: image acquisition (imaging modalities), central archival system, image visualization systems, and network infrastructure. Imaging modalities typically include CT, ultra-sonography, nuclear medicine, and MRI. However, it is possible to extend the functionalities of PACS beyond radiology, including, for example, echocardiography; hematology; examinations of the ear, nose, and throat; and dermatological imaging [129]. The central storage device (archive) stores images, and in some cases, reports, measurements, and other information that reside with the images. Image visualization is performed at the reading workstations. The reading workstation is where the radiologist reviews the patient's study and formulates the diagnosis. The goals of PACS infrastructure are to provide the necessary framework to integrate distributed and heterogeneous imaging systems; provide intelligent database management of all radiology-related information; arrange efficient means of viewing, analyzing, and documenting study results; and furnish a method for effectively communicating study results to the referring physician [130]. The bandwidth requirements for PACS network infrastructure are high because all data must travel over the network at the time of a request. A minimum of 100 Mbps to the desktop with giga-bit backbone/backplane processing speeds are required [131]. Furthermore, because a PACS is required to be continuously available, a load-balanced and fully-redundant meshed network is recommended.

The concept of PACS has been rapidly evolving into the best technological option for tasks such as medical image transmission, archiving, and display, thus forming, along with the RIS and HIS, a basis for a filmless radiology service [132]. Filmless radiology uses a high-bandwidth network that transports digital images among electronic systems that acquire, archive, communicate, and display the images. The goal of implementing a filmless radiology service is to provide enhancement in the form of greater accessibility, integration of images and reports, application of new techniques for the development of new image acquisition protocols, and new methods for processing and displaying images [132]. Such a service should also create economic and social impact by reducing the expenses associated with storage and handling of conventional radiographs, eliminating waste, and minimizing radiation dose [131].

6.4   THE DICOM STANDARD

Originally named the ACR-NEMA standard, the Digital Imaging and Communications in Medicine (DICOM) standard was introduced to the public for the first time at the 1993 RSNA meeting. The initial goal in developing a standard for transmission of digital images was to enable users to retrieve images and associated information from digital imaging devices in a standard format that would be the same across multiple manufacturers [133]. Standardization through DICOM was essential for the development and implementation of a PACS.

The essence of the DICOM standard is that it prescribes a uniform and well-understood set of rules for communication of digital images (communication is defined here as the interchange of information) [133]. Electronic communication is commonly thought of as being divisible into a set of layers, with each layer performing a defined set of functions. This model of communication as a set of layers is part of an international standard for communication, called the International Standards Organization Open Systems Interconnection (ISO-OSI) Reference Model [134, 135]. The DICOM standard makes use of the layered communication structure [133].

The DICOM standard includes a communication protocol that allows devices manufactured by different companies to communicate with one another, both by exchanging digital images so that they can be created, archived, viewed, or printed, and by exchanging other information, such as patient demographics, exam scheduling, and exam reporting. DICOM-conformant systems from different manufacturers can seamlessly communicate with one another, thus granting the user the option of purchasing a CT scanner from one manufacturer, an ultrasonographic system from another company, and a PACS server from a different supplier [136]. DICOM specifies the types of data that can be sent and the required format of the data. In DICOM, different types of image data, such as CR, DR, or ultrasonographic images, are called DICOM objects. DICOM objects are also known as DICOM image object definitions. Common DICOM objects in medicine include CR images, CT, and secondary capture (SC) images, such as those produced by a film scanner [136, 137].

The information objects of DICOM are used to communicate various images and related data between hardware components and devices. However, such information, on its own, is not adequate to ensure proper and complete operation. In addition to communicating with other components, a device is required to perform additional tasks; for example, a workstation displays information, a printer creates a hard copy, and an archival device stores the data provided. Therefore, DICOM provides standardized services that are used with the information objects. These services are built on a set of elemental services. Because DICOM has both composite and normalized information objects, there are both composite and normalized services. Services are performed using a service element [133, 137].

DICOM builds its more complex services using a set of service elements known as DICOM message service elements (DIMSEs). There are five DIMSEs that are used for composite information objects (DIMSE-C) and six that are used for normalized information objects (DIMSE-N). These DIMSEs fall into the categories of operations (such as 'store', which would cause data to be stored) and notifications (such as 'event report', which would notify a device that a certain event had taken place). These simple DIMSEs are used to build the services expected in a PACS [133, 137].

Communication in DICOM is based on an interplay between the participants. The participants are called DICOM service class users (SCU) and DICOM service class providers (SCP), and each of these participants plays a specific role in a DICOM communication event. It is important to understand that the role of a user or provider can change, and is based on the relationship for a specific transaction. For instance, when a viewing workstation queries a server for a list of studies, the workstation is the user (SCU) and the server is the provider (SCP) of that information. It should be noted that when transferring images, the images are sent from an SCU to an SCP. Therefore, if the workstation subsequently requests an image from the PACS server, the server will act as the service user (SCU) and send the image to the workstation that is acting as the provider (SCP) [136, 137]. Each service class performs a different function. Examples of DICOM service classes include store, print, and query/retrieve. The concepts of DICOM services and objects are used within DICOM in terms of a service object pair (SOP). A SOP is a combination of a DICOM service (store images) and an object (for example, an image). The SOP instance is a unique occurrence of a specific SOP class. A SOP tells the user what service class the modality supports and what types of images (DICOM objects) it handles [133, 137]. For example, the CT storage SOP represents the store command as it is used to exchange a CT image object [136].

The DICOM standard addresses many aspects of image communication and is composed of many different DICOM services, each performing a unique function. Some examples of important DICOM service classes are as follows [136].

- *DICOM Storage Service Class* is an essential service for all image acquisition devices. The DICOM storage service class allows the imaging device to send images to a DICOM server for storage.

- *DICOM Query/Retrieve Service Class* is the service that provides the possibility to search (query) an image server and retrieve images for review or have them sent elsewhere. DICOM does not specify how the database is structured, but it does specify how to ask the database for a list of patients and studies, and how to initiate the transfer of the results to a remote device.

- *DICOM Modality Worklist* is the service class designed to maximize the efficiency of a filmless radiology service. It allows a DICOM device to communicate with the HIS and/or RIS and automates the entry of patient demographic information into an image acquisition device. The users of an acquisition device that supports a DICOM modality worklist are presented with a list (worklist) of studies that need to be performed on a given day, instead of being inquired to enter manually the demographic data, which is a laborious and error-prone task.

- *DICOM Print*: Although image visualization on digital monitors has became common, many hospitals still use laser printers to print film copies of images. The DICOM print service class enables the use of printers from different vendors and provides a method of accurately printing images so that they appear similar to how they appear on a monitor.

DICOM requirements can vary among facilities, according to individual practices, and the type of imaging device. Determining which DICOM functionality is needed is an essential consideration in purchasing imaging equipment. The DICOM standard is extensive and it is not necessary for a single device or server to implement the entire standard. Therefore, a document is required to accompany each DICOM implementation that specifies which parts of the standard are supported, known as DICOM conformance statements. Conformance statements can be obtained directly from the vendor and are often made available on the vendor's website. Without such a document, claims of DICOM conformance cannot be verified if a problem arises with connectivity [136].

6.5   CLIENT–SERVER CONFIGURATIONS AND RADIOLOGICAL WORKFLOW

In a filmless environment, RIS/PACS integration is the basis for a successful electronic radiology practice, preventing data inconsistency and assuring the integrity of information among the databases involved. Web technology is one way of achieving this integration since it is based on international standards and can provide information distribution. Since these technologies are being incorporated into many aspects of life, there is a fast learning curve for the end user. In addition, web technology is an appropriate basis for integrating systems based on a client–server model. Figure 6.1 presents a schematic representation of a client–server configuration environment. In this model, a server system handles all complex processing, and communication with the client system is carried out using a standard protocol. Additionally, the interface for data and image display can be standardized and used throughout an enterprise using a web browser. Thus, client and server systems can be managed independently, and data transmission is controlled by an operational system, providing better guarantees in terms of efficiency and stability [126].

Azevedo-Marques et al. [126] present a RIS/PACS integration solution developed at the Radiology Facility, Medical Center, School of Medicine, University of São Paulo, in Ribeirão Preto (HCFMRP), São Paulo, Brazil. The main goal of the project was to build a filmless radiology environment at HCFMRP. The RIS-HCFMRP part of the solution was developed to work under a client–server model, with a relational data model and friendly graphical user interface. Currently, all RIS and HIS data are stored in an Oracle database, and front-end desktop interfaces are implemented using Delphi (a visual programming language). The main RIS modules are patient report generation, patient report search, and management. Access to these modules is defined by the rights assigned to each user, who must be registered in the system with a username and a password [138].

The PACS-HCFMRP part of the solution is composed of software and hardware that form a system supporting several services, such as a DICOM server, the DICOM (storage) server for image archival, the integration interface with the RIS/HIS server, and the web server. The DICOM server used for the PACS implementation is the Central Test Node (CTN) (http://www.erl.wustl.edu/DICOM/ctn.html), which is open-source software developed by the Electronic Radiology Laboratory at the Mallinckrodt Institute of Radiology, to demonstrate the efficacy of the DICOM standard. It uses a relational database to store its configuration and data concerning the acquired images. The configuration structure contains a table known as the "Application Entity" to register all items of equipment that interact in the image communication process. CTN also supplies services or programs that run on both the servers' and clients' sides in order to assure accurate communication. The main service, which handles the acquisition and storage of DICOM images, is the archive server. It receives all the DICOM messages, reads the header of the file (message) it is receiving, and saves the relevant data in the database tables. These data consist of

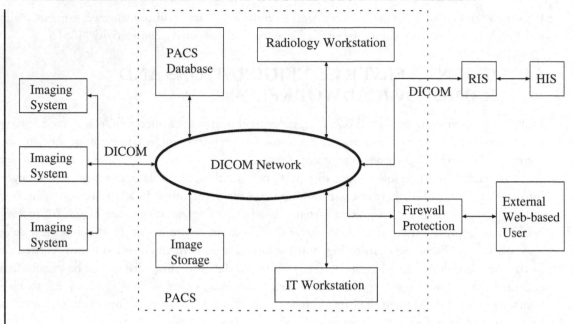

Figure 6.1: Schematic representation of a client–server configuration environment. IT = Information technology.

information about the patient, the exam, and a pointer to the location where the original DICOM file is archived in the server.

A fundamental aspect of the workflow in a digital or filmless radiology environment is the guarantee of the consistency of the data transmitted from one component to the next within the chain of events in the process dynamics. In order to guarantee such consistency, the distribution of information is carried out according to a hierarchical structure which relies on a top-down distribution, that is, the information is propagated from a more general information system (HIS) to an intermediate system (RIS) until it reaches a more specific system (PACS) [139]. Usually, the communication between the RIS and PACS is accomplished by the DICOM modality worklist service. Thus, when a new exam is generated by an imaging modality equipment, some information is inserted into the DICOM header, including the accession number, which is a unique identification generated by the RIS for the specific exam. Images that are generated with the exam information in the DICOM header are then enveloped in a single DICOM file, which is sent to the DICOM server through the hospital network. When the DICOM server receives the file, it extracts the patient and exam information and archives them in its database tables. The acquired original file containing all of the information and the images of the exam is archived in a specific folder that is created in the archiving server under the same accession number. Patient and exam information is shared on a system-wide basis, allowing data integrity and consistency in the RIS/PACS system and assuring reliability to the link of demographic information with the generated images [126].

After performing the preliminary steps described above, it is possible to complete RIS/PACS integration in an effective manner. In order to achieve this, Azevedo-Marques et al. [126] used web technology, with access to and distribution of information by means of a Hyper Text Markup Language (HTML) and Active Server Pages (ASP) application, running on an Internet Information Server (IIS), and connected to two databases via ActiveX Data Objects (ADO). The link between exam reports and images is performed in real time by the hospital network. During a web report search process, a window with a user-friendly interface is opened for the physician who wishes to visualize a specific patient's exam; the window requests the physician's username and password. The system offers a few options, for example, the electronic report search. By selecting this option, a new window is opened in which the user may enter a patient's register number or ask to recover all exams of a specific kind within a certain period. Once the patient and exam are selected, a new window is opened, showing the exam report and information regarding the radiologists who interpreted the exam. The system then performs a link with the RIS/PACS and checks if there are any online images for the same exam in the CTN (DICOM server) databases. If so, a link at the bottom of the report window allows the user to retrieve the images and opens a new window initializing a Java applet for display of the DICOM images. This Java applet, named DICOM Viewer, is open-source software developed and offered by the Iwata Laboratory at the Nagoya Institute of Technology, Japan (http://mars.elcom.nitech.ac.jp/dicom/index-e.html), with the goal of visualizing DICOM images by means of a web browser.

The approach proposed by Azevedo-Marques et al. [126] can be expanded using many other packages of open-source or license-free (freeware) DICOM conformance software. Salomão and Azevedo-Marques [139] presented a model for integration of CAD algorithms into PACS workflow. In their study, a computational testing environment based on the DICOM conQuest server (http://ingenium.home.xs4all.nl/dicom.html) and on the DICOM client and medical image viewer K-PACS (http://www.k-pacs.net/) was built to validate the model. Barra et al. [140] presented a study on freeware medical image viewers. The objective of their study was to search the internet for freeware medical image viewers capable of running as a PACS client, and to evaluate their main functions as well as the feasibility of their use on personal computers. The authors found about 70 image viewers (software); among them, 11 were able to run as PACS clients. Six viewers were selected for analysis based on 16 functions selected by the authors. The authors concluded that although several freeware applications are available, none of them is complete. It is up to the users to analyze and select the software that best suits their needs. Furthermore, the use of freeware image viewers is entirely feasible in the radiologists' daily routine. Besides image viewers, the list of free applications for medical informatics is much more comprehensive [141], including HIS, such as Medical-BR, a HIS with Electronic Health Record (EHR) available at the Brazilian Public Software Portal (http://www.softwarepublico.gov.br/); complex PACS, such as conQuest, PacsOne (http://www.pacsone.net/solutions.htm), OFFIS DCMTK (http://dicom.offis.de/); and the current version of CTN, included in the Linux DebianMed package (http://neuro.debian.net/pkgs/ctn.html) [141].

6.6   INTEGRATION OF CBIR, CAD, PACS, RIS, AND HIS

Radiologists are routinely confronted with clinical tasks that call for quantitative analysis to improve the accuracy and efficiency of image interpretation. CAD can be defined as the diagnosis made by a radiologist using the results of quantitative image processing and analysis from a computer as a second opinion. The second opinion concept is important, because, in the CAD approach, radiologists make the final decisions and the potential gain is due to the synergistic effect obtained by combining the radiologist's competence and the computer's objective capabilities [2]. Although the common workflow in clinical practice still includes CAD as a stand-alone workstation, with the widespread adoption of PACS, there is a promising tendency to integrate CAD systems into the PACS environment [2, 142–145]. This approach appears to hold promise to improve the efficiency of diagnostic examinations in routine clinical work [142, 143]. From the workflow point of view, PACS infrastructure can be used to support query/retrieve tasks and provide the user with images and related patient data based on CAD results that can be directly viewed at a PACS workstation [143]. Furthermore, from an image processing point of view, the vast majority of images in PACS can be considered to be currently "sleeping" [2], and it would be possible to use the quantitative output and results of analysis of CAD for indexing and CBIR [8]. The usefulness of CBIR to assist radiologists in tasks of differential diagnosis has been demonstrated based on observer performance studies [13, 16, 146].

Over the last ten years, solutions have been proposed for CAD-PACS and CBIR-PACS integration and implemented by CAD companies and research teams, by applying DICOM functionalities, DICOM structured reporting (DICOM-SR), and Integrating the Healthcare Enterprise (IHE) workflow profiles [142–144]. Figure 6.2 presents a framework for CBIR-PACS integration. An initial and general approach for CBIR-PACS integration was proposed by Bueno et al. [147] through a system called cbPACS (content-based PACS), which included CBIR resources embedded into the image database manager of the PACS. The proposed structure has four main modules: a system for image acquisition which is responsible for receiving and managing images in the DICOM 3.0 format; a system for image processing which performs all the image manipulation and the extraction of the features used for image indexing and retrieval; a database server that is a DBMS extended to support images as a native data type and responds to similarity queries; and a web server that is responsible for management of the flow of information between the database server and the client units as well as following the applicable data protection and privacy policies. Based on the results obtained from a set of tasks focused on retrieval of similar images, Bueno et al. [147] concluded that the techniques developed and implemented in cbPACS will be able to advance the cause of integration of CBIR into the PACS environment.

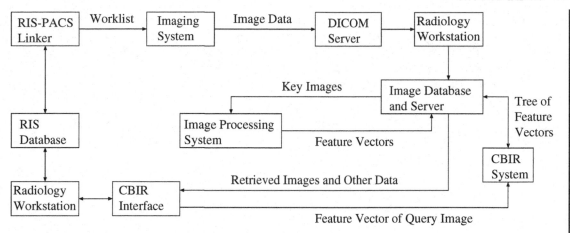

Figure 6.2: Framework for CBIR-PACS integration.

6.7 REMARKS

The integration of the several information and communication systems that make up a radiology service, jointly with PACS, based on the adoption of standards such as DICOM, allows the creation of a filmless digital radiology environment. The integration of a CBIR system into the workflow in a filmless radiology department or clinic makes it possible to retrieve similar images based on a selected query example and, from those images, to retrieve the associated clinical information. In that sense, the integration of CBIR, PACS, RIS, and HIS can be understood as an evidence-based computational tool that can support the diagnostic decision task or as a specific kind of CAD system based on the retrieval of similar cases. Techniques to achieve the above, as described in the present book, could contribute to improved efficiency in diagnostic imaging and better prognosis for patients.

Bibliography

[1] K. Doi. Diagnostic imaging over the last 50 years: research and development in medical imaging science and technology. *Physics in Medicine and Biology*, 51(13):R5–R27, June 2006. DOI: 10.1088/0031-9155/51/13/R02 1

[2] K. Doi. Computer-aided diagnosis in medical imaging: historical review, current status and future potential. *Computerized Medical Imaging and Graphics*, 31(4–5):198–211, 2007. DOI: 10.1016/j.compmedimag.2007.02.002 8, 18, 19, 98

[3] R. M. Rangayyan. *Biomedical Image Analysis*. CRC Press, Boca Raton, FL, 2005. 1, 2, 3, 8, 9, 11, 13, 16, 17, 18, 19, 21, 23, 25, 30, 35, 49, 50, 57, 59, 60, 63, 68

[4] A. P. Dhawan. *Medical Image Analysis*. IEEE and Wiley, New York, NY, 2003. 1

[5] H. H. Barrett and K. J. Myers. *Foundations of Image Science*. Wiley, Hoboken, NJ, 2004. 10

[6] W. Huda and R. Slone. *Review of Radiologic Physics*. Williams and Wilkins, Baltimore, MD, 1995. 2, 8, 9

[7] R. M. Rangayyan, B. Acha, and C. Serrano. *Color Image Processing with Biomedical Applications*. SPIE, Bellingham, WA, 2011. DOI: 10.1117/3.887920 3

[8] H. Müller, N. Michoux, D. Bandon, and A. Geissbühler. A review of content-based image retrieval systems in medical applications: clinical benefits and future directions. *International Journal of Medical Informatics*, 73:1–23, May 2004. DOI: 10.1016/j.ijmedinf.2003.11.024 3, 4, 5, 68, 98

[9] I. El-Naqa, Y. Yang, N. P. Galatsanos, R. M. Nishikawa, and M. N. Wernick. A similarity learning approach to content-based image retrieval: application to digital mammography. *IEEE Transactions on Medical Imaging*, 23:1233–1244, 2004. DOI: 10.1109/TMI.2004.834601 4, 87, 89

[10] H. Alto, R. M. Rangayyan, and J. E. L. Desautels. Content-based retrieval and analysis of mammographic masses. *Journal of Electronic Imaging*, 14(2):023016:1–17, 2005. DOI: 10.1117/1.1902996 4, 8, 16, 17, 51, 87, 89

[11] S. K. Kinoshita, P. M. Azevedo-Marques, R. R. Pereira Jr., J. A. H. Rodrigues, and R. M. Rangayyan. Content-based retrieval of mammograms using visual features related to breast density patterns. *Journal of Digital Imaging*, 20(2):172–190, June 2007. DOI: 10.1007/s10278-007-9004-0 4, 30, 45, 59, 61, 67, 71

[12] P. M. Azevedo-Marques, N. A. Rosa, A. J. M. Traina, C. Traina Jr., S. K. Kinoshita, and R. M. Rangayyan. Reducing the semantic gap in content-based image retrieval in mammography with relevance feedback and inclusion of expert knowledge. *International Journal of Computer Assisted Radiology and Surgery*, 3(1–2):123–130, 2008. DOI: 10.1007/s11548-008-0154-4 4, 5, 19, 67, 78, 80, 83, 84

[13] Q. Li, F. Li, J. Shiraishi, S. Katsuragawa, S. Sone, and K. Doi. Investigation of new psychophysical measures for evaluation of similar images on thoracic computed tomography for distinction between benign and malignant nodules. *Medical Physics*, 30:2584–2593, 2003. DOI: 10.1118/1.1605351 8, 19, 98

[14] G. D. Tourassi, N. H. Eltonsy, J. Graham, C. E. Floyd, and A. S. Elmaghraby. Feature and knowledge based analysis for reduction of false positives in the computerized detection of masses in screening mammography. In *Proceedings of the 27th Annual International Conference of the IEEE Engineering in Medicine and Biology Society (CD-ROM)*, pages 6524–6527, 2005. DOI: 10.1109/IEMBS.2005.1615994 4

[15] R. M. Haralick, K. Shanmugam, and I. Dinstein. Textural features for image classification. *IEEE Transactions on Systems, Man, Cybernetics*, 3(6):610–622, 1973. DOI: 10.1109/TSMC.1973.4309314 4, 51, 53, 57

[16] C. Muramatsu, Q. Li, K. Suzuki, R. A. Schmidt, J. Shiraishi, G. M. Newstead, and K. Doi. Investigation of psychophysical measure for evaluation of similar images for mammographic masses: preliminary results. *Medical Physics*, 32:2295–2304, 2005. DOI: 10.1118/1.1944913 4, 74, 98

[17] E. Y. Tao and J. Sklansky. Analysis of mammograms aided by database of images of calcifications and textures. In *Proceedings of SPIE Medical Imaging 1996: Image Processing*, pages 988–995. SPIE, 1996. DOI: 10.1117/12.237907 4

[18] C. J. Ornes, D. J. Valentino, H. J. Yoon, J. I. Eisenman, and J. Sklansky. Search engine for remote database-aided interpretation of digitized mammograms. In E. L. Siegel and H. K. Huang, editors, *Proceedings of SPIE Medical Imaging 2001 on PACS and Integrated Medical Information Systems: Design and Evaluation*, pages 132–137. SPIE, 2001. 4

[19] I. El-Naqa, Y. Yang, N. P. Galatsanos, and M. N. Wernick. Content-based image retrieval for digital mammography. In *Proceedings of the IEEE International Conference on Image Processing*, pages 141–144, Rochester, NY, 2002. DOI: 10.1109/ICIP.2002.1038924 4

[20] T. Nakagawa, T. Hara, H. Fujita, T. Iwase, and T. Endo. Image retrieval system of mammographic masses by using local pattern matching technique. In H.-O. Peitgen, editor, *Proceedings of the 6th International Workshop on Digital Mammography: IWDM 2002*, pages 562–565, Bremen, Germany, January 2002. Springer. 4

[21] T. Nakagawa, T. Hara, H. Fujita, T. Iwase, and T. Endo. Development of a computer-aided sketch system for mammograms. In H.-O. Peitgen, editor, *Proceedings of the 6th International Workshop on Digital Mammography: IWDM 2002*, pages 581–583, Bremen, Germany, January 2002. Springer. 4

[22] T. M. Deserno, S. Antani, and R. Long. Ontology of gaps in content-based image retrieval. *Journal of Digital Imaging*, 22(2):202–215, 2009. DOI: 10.1007/s10278-007-9092-x 5

[23] R. A. Baeza-Yates and B. A. Ribeiro-Neto. *Modern Information Retrieval.* ACM Press, New York, NY, 1999. 5

[24] M. Ortega-Binderberger and S. Mehrotra. Relevance feedback in multimedia databases. In B. Furht and O. Marques, editors, *Handbook of Video Databases: Design and Applications*, pages 511–536. CRC Press, Boca Raton, FL, 2003. 5

[25] X. S. Zhou and T. S. Huang. Relevance feedback in image retrieval: A comprehensive review. *Multimedia Systems*, 8:536–544, 2003. DOI: 10.1007/s00530-002-0070-3 5

[26] J. S. Spratt and J. A. Spratt. Growth rates. In W. L. Donegan and J. S. Spratt, editors, *Cancer of the Breast*, chapter 10, pages 270–302. Saunders, Philadelphia, PA, 3rd edition, 1988. 7

[27] M. J. Homer. *Mammographic Interpretation: A Practical Approach.* McGraw-Hill, New York, NY, 2nd edition, 1997. 7, 13, 51

[28] L. W. Basset and R. H. Gold, editors. *Breast Cancer Detection: Mammography and Other Methods in Breast Imaging.* Grune & Stratton, Orlando, FL, 2nd edition, 1987.

[29] G. Cardenosa. *Breast Imaging Companion.* Lippincott-Raven, Philadelphia, PA, 1997.

[30] S. H. Heywang-Köbrunner, I. Schreer, and D. D. Dershaw. *Diagnostic Breast Imaging: Mammography, Sonography, Magnetic Resonance Imaging, and Interventional Procedures.* Thieme Medical Publishers, New York, NY, 1997.

[31] A. G. Haus. Recent trends in screen-film mammography: Technical factors and radiation dose. In S. Brünner and B. Langfeldt, editors, *Recent Results in Cancer Research*, volume 105, pages 37–51. Springer-Verlag, Berlin, Germany, 1987. 7

[32] J. N. Wolfe. Risk for breast cancer development determined by mammographic parenchymal pattern. *Cancer*, 37:2486–2492, 1976.
DOI: 10.1002/1097-0142(197605)37:5%3C2486::AID-CNCR2820370542%3E3.0.CO;2-8
7, 13

[33] N. F. Boyd, J. W. Byng, R. A. Jong, E. K. Fishell, L. E. Little, A. B. Miller, G. A. Lockwood, D. L. Tritchler, and M. J. Yaffe. Quantitative classification of mammographic densities and breast cancer risk: results from the Canadian National Breast Screening Study. *Journal of the National Cancer Institute*, 87(9):670–675, May 1995. DOI: 10.1093/jnci/87.9.670 7

[34] American College of Radiology (ACR). *Illustrated Breast Imaging — Reporting and Data System (BI-RADS)*. American College of Radiology, Reston, VA, fourth edition, 2003. 7, 13, 18, 51

[35] C. Zhou, H. P. Chan, N. Petrick, M. A. Helvie, M. M. Goodsitt, B. Sahiner, and L. M. Hadjiiski. Computerized image analysis: Estimation of breast density on mammograms. *Medical Physics*, 28(6):1056–1069, 2001. DOI: 10.1118/1.1376640 8, 19, 25

[36] Y. Rui, T. S. Huang, M. Ortega, and S. Mehrotra. Relevance feedback: a power tool for interactive content-based image retrieval. *IEEE Transactions on Circuits and Systems: Video Technology*, 8:644–655, 1998. DOI: 10.1109/76.718510 19

[37] R. M. Rangayyan, F. J. Ayres, and J. E. L. Desautels. A review of computer-aided diagnosis of breast cancer: Toward the detection of subtle signs. *Journal of the Franklin Institute*, 344:312–348, 2007. DOI: 10.1016/j.jfranklin.2006.09.003 13, 18

[38] J. Tang, R. M. Rangayyan, J. Xu, I. E. Naqa, and Y. Yang. Computer-aided detection and diagnosis of breast cancer with mammography: Recent advances. *IEEE Transactions on Information Technology in Biomedicine*, 13(2):236–251, March 2009. DOI: 10.1109/TITB.2008.2009441 8, 13, 18

[39] R. A. Robb. X-ray computed tomography: An engineering synthesis of mulitscientific principles. *CRC Critical Reviews in Biomedical Engineering*, 7:264–333, 1982. DOI: 10.1146/annurev.bb.11.060182.001141 8, 35

[40] Z. H. Cho, J. P. Jones, and M. Singh. *Foundations of Medical Imaging*. Wiley, New York, NY, 1993. 8, 10

[41] A. Macovski. *Medical Imaging Systems*. Prentice-Hall, Englewood Cliffs, NJ, 1983. 8, 9, 10

[42] E. D. Pisano, C. Gatsonis, E. Hendrick, M. Yaffe, J. K. Baum, S. Acharyya, E. F. Conant, L. L. Fajardo, L. Bassett, C. D'Orsi, R. Jong, and M. Rebner. Diagnostic performance of digital versus film mammography for breast-cancer screening. *New England Journal of Medicine*, 353(17):1773–1783, October 2005. DOI: 10.1056/NEJMoa052911 10

[43] E. D. Pisano. Current status of full-field digital mammography. *Radiology*, 214:26–28, 2000. DOI: 10.1016/S1076-6332(00)80478-X

[44] M. J. Yaffe. Development of full field digital mammography. In N. Karssemeijer, M. Thijssen, J. Hendriks, and L. van Erning, editors, *Proceedings of the 4th International Workshop on Digital Mammography*, pages 3–10, Nijmegen, The Netherlands, June 1998. DOI: 10.1007/978-94-011-5318-8

[45] L. Cheung, R. Bird, A. Chitkara, A. Rego, C. Rodriguez, and J. Yuen. Initial operating and clinical results of a full field mammography system. In N. Karssemeijer, M. Thijssen, J. Hendriks, and L. van Erning, editors, *Proceedings of the 4th International Workshop on Digital Mammography*, pages 11–18, Nijmegen, The Netherlands, June 1998. DOI: 10.1007/978-94-011-5318-8

[46] A. D. A. Maidment, R. Fahrig, and M. J. Yaffe. Dynamic range requirements in digital mammography. *Medical Physics*, 20(6):1621–1633, 1993. DOI: 10.1118/1.596949 10

[47] I. Andersson. Mammography in clinical practice. *Medical Radiography and Photography*, 62(2):1–41, 1986. 10

[48] E. A. Sickles and W. N. Weber. High-contrast mammography with a moving grid: Assessment of clinical utility. *American Journal of Radiology*, 146:1137–1139, 1986. 10

[49] E. A. Sickles. The role of magnification technique in modern mammography. In S. Brünner and B. Langfeldt, editors, *Recent Results in Cancer Research*, volume 105, pages 19–24. Springer-Verlag, Berlin, Germany, 1987. 10

[50] G. Ursin, L. Hovanessian-Larsen, Y. Parisky, M. C. Pike, and A. Wu. Greatly increased occurrence of breast cancers in areas of mammographically dense tissue. *Breast Cancer Research*, 7:R605–R608, 2005. DOI: 10.1186/bcr1260 13

[51] G. Ursin and S. A. Qureshi. Mammographic density – a useful biomarker for breast cancer risk in epidemiologic studies. *Norsk Epidemiologi*, 19(1):59–68, 2009. 13

[52] N. Karssemeijer. Automated classification of parenchymal patterns in mammograms. *Physics in Medicine and Biology*, 43(2):365–378, 1998. DOI: 10.1088/0031-9155/43/2/011 13, 25

[53] J. W. Byng, N. F. Boyd, E. Fishell, R. A. Jong, and M. J. Yaffe. Automated analysis of mammographic densities. *Physics in Medicine and Biology*, 41:909–923, 1996. DOI: 10.1088/0031-9155/41/5/007 51

[54] J. N. Wolfe. Breast parenchymal patterns and their changes with age. *Radiology*, 121:545–552, 1976.

[55] J. W. Byng, N. F. Boyd, E. Fishell, R. A. Jong, and M. J. Yaffe. The quantitative analysis of mammographic densities. *Physics in Medicine and Biology*, 39:1629–1638, 1994. DOI: 10.1088/0031-9155/39/10/008

[56] P. G. Tahoces, J. Correa, M. Souto, L. Gómez, and J. J. Vidal. Computer-assisted diagnosis: the classification of mammographic breast parenchymal patterns. *Physics in Medicine and Biology*, 40:103–117, 1995. DOI: 10.1088/0031-9155/40/1/010

[57] Z. Huo, M. L. Giger, W. Zhong, and O. I. Olopade. Analysis of relative contributions of mammographic features and age to breast cancer risk prediction. In M. J. Yaffe, editor, *Proceedings of the 5th International Workshop on Digital Mammography*, pages 732–736, Toronto, ON, Canada, June 2000.

[58] R. Sivaramakrishna, N. A. Obuchowski, W. A. Chilcote, and K. A. Powell. Automatic segmentation of mammographic density. *Academic Radiology*, 8(3):250–256, 2001. DOI: 10.1016/S1076-6332(03)80534-2

[59] P. K. Saha, J. K. Udupa, E. F. Conant, D. P. Chakraborty, and D. Sullivan. Breast tissue density quantification via digitized mammograms. *IEEE Transactions on Medical Imaging*, 20(8):792–803, 2001. DOI: 10.1109/42.938247 13

[60] H. Zonderland. The radiology assistant, accessed September 2012. `http://www.radiologyassistant.nl/en/4349108442109#a43db92b2e8101.` 13

[61] L. Shen, R. M. Rangayyan, and J. E. L. Desautels. Application of shape analysis to mammographic calcifications. *IEEE Transactions on Medical Imaging*, 13(2):263–274, June 1994. 16

[62] L. Shen, R. M. Rangayyan, and J. E. L. Desautels. Detection and classification of mammographic calcifications. *International Journal of Pattern Recognition and Artificial Intelligence*, 7(6):1403–1416, 1993. DOI: 10.1142/S0218001493000686 16

[63] C. Serrano, B. Acha, R. M. Rangayyan, and J. E. L. Desautels. Detection of microcalcifications in mammograms using error of prediction and statistical measures. *Journal of Electronic Imaging*, 18(1):013011:1–10, 2009. 16

[64] R. M. Rangayyan, N. R. Mudigonda, and J. E. L. Desautels. Boundary modelling and shape analysis methods for classification of mammographic masses. *Medical and Biological Engineering and Computing*, 38:487–496, 2000. DOI: 10.1007/BF02345742 16, 17

[65] R. M. Rangayyan and T. M. Nguyen. Fractal analysis of contours of breast masses in mammograms. *Journal of Digital Imaging*, 20(3):223–237, September 2007. DOI: 10.1007/s10278-006-0860-9 16, 17

[66] N. R. Mudigonda, R. M. Rangayyan, and J. E. L. Desautels. Detection of breast masses in mammograms by density slicing and texture flow-field analysis. *IEEE Transactions on Medical Imaging*, 20(12):1215–1227, 2001. DOI: 10.1109/42.974917 17, 51

[67] J. S. Lee. Digital image enhancement and noise filtering by use of local statistics. *IEEE Transactions on Pattern Analysis and Machine Intelligence*, PAMI-2:165–168, March 1980. DOI: 10.1109/TPAMI.1980.4766994 23

[68] P. Perona and J. Malik. Scale-space and edge detection using anisotropic diffusion. *IEEE Transactions on Pattern Analysis and Machine Intelligence*, 12(7):629–639, 1990. DOI: 10.1109/34.56205 24

[69] F. Catté, P. L. Lons, J. M. Morel, and T. Coll. Image selective smoothing and edge detection by nonlinear diffusion–I. *SIAM Journal of Numerical Analysis*, 29(1):182–193, 1992. DOI: 10.1137/0729012 24

[70] F. Catté, P. L. Lons, J. M. Morel, and T. Coll. Image selective smoothing and edge detection by nonlinear diffusion–II. *SIAM Journal of Numerical Analysis*, 29(1):845–866, 1992. DOI: 10.1137/0729012 24

[71] C. A. Segall and S. T. Acton. Morphological anisotropic diffusion. In *IEEE International Conference on Image Processing*, pages 348–351, Santa Barbara, CA, October 1997. DOI: 10.1109/ICIP.1997.632112 25

[72] Z. Huo, M. L. Giger, D. E. Wolverton, W. Zhong, and S. Cumming. Computerized analysis of mammographic parenchymal patterns for breast cancer risk assessment: feature selection. *Medical Physics*, 27(1):4–12, 2000. DOI: 10.1118/1.598851 25

[73] S. K. Kinoshita, P. M. Azevedo-Marques, A. F. Frère, H. R. C. Marana, R. J. Ferrari, and R. L. Villela. Comparative analysis of shape and texture features in classification of breast masses in digitized mammograms. In *Proceedings of SPIE Volume 3979, Medical Imaging 2000*, pages 872–879, 2000.

[74] F. F. Yin, M. L. Giger, K. Doi, C. E. Metz, C. J. Vyborny, and R.A. Schmidt. Computerized detection of masses in digital mammograms: Analysis of bilateral subtraction images. *Medical Physics*, 18(5):955–963, 1991. DOI: 10.1118/1.596610 25

[75] R. J. Ferrari, R. M. Rangayyan, J. E. L. Desautels, and A. F. Frère. Analysis of asymmetry in mammograms via directional filtering with Gabor wavelets. *IEEE Transactions on Medical Imaging*, 20(9):953–964, 2001. DOI: 10.1109/42.952732 25

[76] R. M. Rangayyan, R. J. Ferrari, and A. F. Frère. Analysis of bilateral asymmetry in mammograms using directional, morphological and density features. *Journal of Electronic Imaging*, 16(1):013003: 1–12, 2007. DOI: 10.1117/1.2712461 25

[77] R. J. Ferrari, R. M. Rangayyan, J. E. L. Desautels, R. A. Borges, and A. F. Frère. Automatic identification of the pectoral muscle in mammograms. *IEEE Transactions on Medical Imaging*, 23:232–245, 2004. DOI: 10.1109/TMI.2003.823062 25, 35, 41

[78] R. J. Ferrari, R. M. Rangayyan, J. E. L. Desautels, R. A. Borges, and A. F. Frère. Identification of the breast boundary in mammograms using active contour models. *Medical and Biological Engineering and Computing*, 42:201–208, 2004. DOI: 10.1007/BF02344632

[79] R. J. Ferrari, R. M. Rangayyan, R. A. Borges, and A. F. Frère. Segmentation of the fibro-glandular disc in mammograms using Gaussian mixture modeling. *Medical and Biological Engineering and Computing*, 42:378–387, 2004. DOI: 10.1007/BF02344714 25, 30

[80] A. J. Méndez, P. G. Tahoces, M. J. Lado, M. Souto, J. L. Correa, and J. J. Vidal. Automatic detection of breast border and nipple in digital mammograms. *Computer Methods and Programs in Biomedicine*, 49:253–262, 1996. DOI: 10.1016/0169-2607(96)01724-5 25, 27

[81] R. M. Rangayyan and W. A. Rolston. Directional image analysis with the Hough and Radon transforms. *Journal of the Indian Institute of Science*, 78:3–16, 1998. 27, 35

[82] N. Srinivasa, K. R. Ramakrishnan, and K. Rajgopal. Detection of edges from projections. *IEEE Transactions on Medical Imaging*, 11(1):76–80, 1992. DOI: 10.1109/42.126913 27, 35

[83] S. K. Kinoshita, P. M. Azevedo-Marques, R. R. Pereira Jr., J. A. H. Rodrigues, and R. M. Rangayyan. Radon-domain detection of the nipple and the pectoral muscle in mammograms. *Journal of Digital Imaging*, 21(1):37–49, March 2007. DOI: 10.1007/s10278-007-9035-6 27, 30, 35, 41

[84] R. Chandrasekhar and Y. Attikiouzel. A simple method for automatically locating the nipple on mammograms. *IEEE Transactions on Medical Imaging*, 16(5):483–494, 1997. DOI: 10.1109/42.640738 27

[85] J. N. Kapur, P. K. Sahoo, and A. K. C. Wong. A new method for grey-level picture thresholding using the entropy of the histogram. *Computer Vision, Graphics, and Image Processing*, 99:273–285, 1985. DOI: 10.1016/0165-1684(80)90020-1 27, 29, 30

[86] W. Tsai. Moment-preserving thresholding: a new approach. *Computer Vision, Graphics, and Image Processing*, 99:377–393, 1985. 27, 30

[87] N. Otsu. A threshold selection method from gray-level histograms. *IEEE Transactions on Systems, Man, and Cybernetics*, 9(1):62–66, 1979. DOI: 10.1109/TSMC.1979.4310076 27, 30

[88] T. Ridler and S. Calvard. Picture thresholding using an iterative selection method. *IEEE Transactions on Systems, Man, and Cybernetics*, 8:630–632, 1978. DOI: 10.1109/TSMC.1978.4310039 27, 29

[89] S. S. Reddi, S. F. Rudin, and H. R. Keshavan. An optimal multiple threshold scheme for image segmentation. *IEEE Transactions on Systems, Man, and Cybernetics*, 14:661–665, 1984. DOI: 10.1109/21.57290 27

[90] P. K. Sahoo, S. Soltani, and A. K. C. Wong. A survey of thresholding techniques. *Computer Vision, Graphics, and Image Processing*, 41:233–260, 1988. DOI: 10.1016/0734-189X(88)90022-9 27

[91] R. C. Gonzalez and R. E. Woods. *Digital Image Processing*. Prentice-Hall, Upper Saddle River, NJ, 2nd edition, 2002. 30, 41, 60, 63

[92] E. R. Dougherty. *An Introduction to Morphological Image Processing*. SPIE, Bellingham, WA, 1992. 30, 41

[93] S. K. Kinoshita. *Atributos Visuais para Recuperação Baseada em Conteúdo de Imagens Mamográ-ficas*. PhD thesis, School of Engineering of São Carlos, University of São Paulo, São Carlos, São Paulo, Brazil, 2004. 30, 33, 47

[94] A. Rosenfeld and A. C. Kak. *Digital Picture Processing*. Academic Press, London, UK, 2nd edition, 1982. 30, 35, 49

[95] M. P. Sampat, G. J. Whitman, M. K. Markey, and A. C. Bovik. Evidence based detection of spiculated masses and architectural distortion. In J. M. Fitzpatrick and J. M. Reinhardt, editors, *Proceedings of SPIE Medical Imaging 2005: Image Processing*, volume 5747, pages 26–37, San Diego, CA, April 2005. 35

[96] A. C. G. Martins and R. M. Rangayyan. Complex cepstral filtering of images and echo removal in the Radon domain. *Pattern Recognition*, 30(11):1931–1938, 1997. DOI: 10.1016/S0031-3203(97)00010-1 35

[97] A. C. G. Martins and R. M. Rangayyan. Texture element extraction via cepstral filtering in the Radon domain. *IETE Journal of Research (India)*, 48(3,4):143–150, 2002. 35

[98] J. Canny. A computational approach to edge detection. *IEEE Transactions on Pattern Analysis and Machine Intelligence*, 8(6):679–698, 1986. DOI: 10.1109/TPAMI.1986.4767851 35

[99] E. L. Hall. *Computer Image Processing and Recognition*. Academic, New York, NY, 1979. 49, 63

[100] C. E. Shannon. A mathematical theory of communication. *Bell System Technical Journal*, 27:379–423, 623–656, 1948. DOI: 10.1145/584091.584093

[101] C. E. Shannon. Communication in the presence of noise. *Proceedings of the IRE*, 37:10–21, 1949. DOI: 10.1109/JRPROC.1949.232969 49

[102] A. K. Jain. *Fundamentals of Digital Image Processing*. Prentice-Hall, Englewood Cliffs, NJ, 2nd edition, 1989. 50

[103] R. M. Rangayyan, N. M. El-Faramawy, J. E. L. Desautels, and O. A. Alim. Measures of acutance and shape for classification of breast tumors. *IEEE Transactions on Medical Imaging*, 16(6):799–810, December 1997. DOI: 10.1109/42.650876 51, 60, 61

[104] B. S. Sahiner, H. P. Chan, N. Petrick, M. A. Helvie, and L. M. Hadjiiski. Improvement of mammographic mass characterization using spiculation measures and morphological features. *Medical Physics*, 28(7):1455–1465, 2001. DOI: 10.1118/1.1381548

[105] N. R. Mudigonda, R. M. Rangayyan, and J. E. L. Desautels. Gradient and texture analysis for the classification of mammographic masses. *IEEE Transactions on Medical Imaging*, 19(10):1032–1043, 2000. DOI: 10.1109/42.887618 51

[106] G. N. Lee, T. Hara, and H. Fujita. Classifying masses as benign or malignant based on cooccurrence matrix textures: A comparison study of different gray level quantizations. In *Digital Mammography: Lecture Notes in Computer Science, Volume 4046/2006*, pages 332–339, Berlin, Germany, 2006. Springer. DOI: 10.1007/11783237_45 51

[107] R. M. Rangayyan, T. M. Nguyen, F. J. Ayres, and A. K. Nandi. Effect of pixel resolution on texture features of breast masses in mammograms. *Journal of Digital Imaging*, 23(5):547–553, October 2010. DOI: 10.1007/s10278-009-9238-0 53

[108] T. M. Cabral and R. M. Rangayyan. *Fractal Analysis of Breast Masses in Mammograms*. Morgan & Claypool, 2012. DOI: 10.2200/S00453ED1V01Y201210BME046 53

[109] E. R. Dougherty, J. T. Newell, and J. B. Pelz. Morphological texture-based maximum-likelihood pixel classification based on local granulometric moments. *Pattern Recognition*, 25(10):1181–1198, 1992. DOI: 10.1016/0031-3203(92)90020-J 57

[110] L. Gupta and M. D. Srinath. Contour sequence moments for the classification of closed planar shapes. *Pattern Recognition*, 20(3):267–272, 1987. DOI: 10.1016/0031-3203(87)90001-X 60

[111] S. A. Dudani, K. J. Breeding, and R. B. McGhee. Aircraft identification by moment invariants. *IEEE Transactions on Computers*, C-26(1):39–45, 1983. DOI: 10.1109/TC.1977.5009272

[112] M. K. Hu. Visual pattern recognition by moment invariants. *IRE Transactions on Information Theory*, IT-8(2):179–187, 1962. DOI: 10.1109/TIT.1962.1057692 60, 61

[113] K. Fukunaga. *Introduction to Statistical Pattern Recognition*. Academic, San Diego, CA, 2 edition, 1990. 62

[114] R. O. Duda, P. E. Hart, and D. G. Stork. *Pattern Classification*. Wiley-Interscience, New York, NY, 2nd edition, 2001. 68

[115] R. J. Nandi, A. K. Nandi, R. M. Rangayyan, and D. Scutt. Classification of breast masses in mammograms using genetic programming and feature selection. *Medical and Biological Engineering and Computing*, 44:683–694, 2006. DOI: 10.1007/s11517-006-0077-6 62

[116] J. T. Tou and R. C. Gonzalez. *Pattern Recognition Principles*. Addison-Wesley, Reading, MA, 1974. 68

[117] T. Kohonen. The self-organizing map. *Proceedings of the IEEE*, 78(9):1464–1480, 1990. DOI: 10.1109/5.58325 69, 71

[118] J. Laaksonen, M. Koskela, S. Laakso, and E. Oja. PicSOM – content-based image retrieval with self-organizing maps. *Pattern Recognition Letters*, 21(13-14):1199–1207, 2000. DOI: 10.1016/S0167-8655(00)00082-9 71

[119] J. J. Rocchio. Relevance feedback in information retrieval. In E. Cliffs, editor, *The SMART Retrieval System: Experiments in Automatic Document Processing*, pages 313–323. Prentice Hall, Englewood Cliffs, NJ, 1971. 80

[120] A. Traina, J. Marques, and C. Traina Jr. Fighting the semantic gap on CBIR systems through new relevance feedback techniques. In *CBMS'06: Proceedings of the 19th IEEE International Symposium on Computer-based Medical Systems*, pages 881–886, Salt Lake City, UT, 2006. DOI: 10.1109/CBMS.2006.88 80

[121] S. Banik, R. M. Rangayyan, and J. E. L. Desautels. Detection of architectural distortion in prior mammograms. *IEEE Transactions on Medical Imaging*, 30(2):279–294, February 2011. DOI: 10.1109/TMI.2010.2076828 88

[122] Q. Guo, J. Shao, and V. F. Ruiz. Characterization and classification of tumor lesions using computerized fractal-based texture analysis and support vector machines in digital mammograms. *International Journal of Computer Assisted Radiology and Surgery*, 4(1):11–25, January 2009. DOI: 10.1007/s11548-008-0276-8

[123] C. B. Caldwell, S. J. Stapleton, D. W. Holdsworth, R. A. Jong, W. J. Weiser, G. Cooke, and M. J. Yaffe. Characterization of mammographic parenchymal pattern by fractal dimension. *Physics in Medicine and Biology*, 35(2):235–247, 1990. DOI: 10.1088/0031-9155/35/2/004 88

[124] H. K. Huang. *PACS and Imaging Informatics: Basic Principles and Applications*. Wiley, New York, NY, 2004. DOI: 10.1002/0471654787 91

[125] E. H. Shortliffe and J. J. Cimino. *Biomedical Informatics: Computer Applications in Health Care and Biomedicine*. Springer, New York, NY, 3rd edition, 2006. 91

[126] P. M. Azevedo-Marques, E. C. Caritá, A. A. Benedicto, and P. R. Sanches. Integrating RIS/PACS: The Web-based solution at University Hospital of Ribeirão Preto, Brazil. *Journal of Digital Imaging*, 17:226–233, 2004. DOI: 10.1007/s10278-004-1018-2 91, 95, 97

[127] G. Wiley. The prophet motive: How PACS was developed and sold, accessed October 2012. www.imagingeconomics.com/issues/articles/2005-05_01.asp. 91

[128] Digital Imaging and Communications in Medicine, Part 1–10, Nema Standard Publication 1992–2000. http://medical.nema.org/. 92

[129] D. Bandon, C. Lovis, A. Geissbühler, and J.-P. Vallée. Enterprise-wide PACS: beyond radiology, an architecture to manage all medical images. *Academic Radiology*, 12:1000–1009, 2005. DOI: 10.1016/j.acra.2005.03.075 92

[130] H. K. Huang and R. K. Taira. Infrastructure design of a picture archiving and communication system. *American Journal of Radiology*, 158:743–749, 1992. 92

[131] A. F. Goldszal, M. H. Bleshman, and R. N. Bryan. Financing a large-scale picture archival and communication system. *Academic Radiology*, 11:96–102, 2004. DOI: 10.1016/S1076-6332(03)00544-0 92

[132] E. L. Siegel. Current state of the art and future trends. In E. L. Siegel and R. M. Kolodner, editors, *Filmless Radiology*, pages 3–20. Springer, New York, NY, 1999. DOI: 10.1007/978-1-4612-1402-1 92

[133] S. C. Horii. Primer on computers and information technology. Part Four: A nontechnical introduction to DICOM. *RadioGraphics*, 17:1297–1309, 1997. 92, 93, 94

[134] A. S. Tanenbaum. *Computer Networks*. Prentice-Hall, Englewood Cliffs, NJ, 2nd edition, 1988. 93

[135] F. Helsall. *Data Communications, Computer Networks, and Open Systems*. Addison-Wesley, Wokingham, UK, 3rd edition, 1988. 93

[136] M. A. Wright, D. Balance, I. D. Robertson, and B. Poteet. Introduction to DICOM for the practicing veterinarian. *Veterinary Radiology & Ultrasound*, 49:S14–S18, 2008. DOI: 10.1111/j.1740-8261.2007.00328.x 93, 94, 95

[137] P. M. Azevedo-Marques and S. C. Salomão. PACS: Sistemas de arquivamento e distribuição de imagens. *Revista Brasileira de Física Médica*, 3:131–139, 2009. 93, 94

[138] P. M. Azevedo-Marques, A. C. Santos, J. Elias, Jr., W. M. Goes, C. R. Castro, and C. S. Trad. Implantação de um sistema de informação em radiologia em hospital universitário. *Radiologia Brasileira*, 33:155–160, 2000. 95

[139] S. C. Salomão and P. M. Azevedo-Marques. Integrating computer-aided diagnosis tools into the picture archiving and communication system. *Radiologia Brasileira*, 44:374–380, 2011. DOI: 10.1590/S0100-39842011000600009 96, 97

[140] F. R. Barra, R. R. Barra, and A. Barra Sobrinho. Freeware medical image viewers: can we rely only on them? *Radiologia Brasileira*, 43:313–318, 2010. 97

[141] L. F. Nobre and A. von Wangenheim. Free software: an option for radiologists? *Radiologia Brasileira*, 43:IX–X, 2010. 97, 98

[142] Z. Zhou, B. J. Liu, and A. H. Le. CAD-PACS integration tool kit based on DICOM secondary capture, structured report and IHE workflow profiles. *Computerized Medical Imaging and Graphics*, 31:198–211, 2007. DOI: 10.1016/j.compmedimag.2007.02.015 98

[143] A. H. T. Le, B. Liu, and H. K. Huang. Integration of computer-aided diagnosis/detection (CAD) results in a PACS environment using CAD-PACS toolkit and DICOM SR. *International Journal of Computer Assisted Radiology and Surgery*, 4:317–329, 2009. DOI: 10.1007/s11548-009-0297-y 98

[144] P. Welter, T. M. Deserno, C. Grouls, and R. W. Günther. Workflow management of content-based image retrieval for CAD support in PACS environments based on IHE. *International Journal of Computer Assisted Radiology and Surgery*, 5:393–400, 2010. DOI: 10.1007/s11548-010-0416-9 98

[145] C. Traina Jr., A. J. M. Traina, M. R. B. Araújo, J. M. Bueno, F. J. T. Chino, H. L. Razente, and P. M. Azevedo-Marques. Using an image-extended relational database to support content-based image retrieval in a PACS. *International Journal of Computer Methods and Programs in Biomedicine*, 80(1):71–83, 2005. DOI: 10.1016/S0169-2607(05)80008-2 98

[146] C. Muramatsu, Q. Li, R. A. Schmidt, K. Suzuki, J. Shiraishi, G. M. Newstead, and K. Doi. Experimental determination of subjective similarity for pairs of clustered microcalcifications on mammograms: observer study results. *Medical Physics*, 33:3460–3468, 2006. DOI: 10.1118/1.2266280 98

[147] J. M. Bueno, F. J. T. Chino, A. J. M. Traina, C. Traina Jr., and P. M. Azevedo-Marques. How to add content-based image retrieval capability in a PACS. In *Proceedings of the IEEE International Conference on Computer Based Medical Systems — CBMS*, pages 321–326, Maribor, Slovenia, 2002. IEEE Computer Society. DOI: 10.1109/CBMS.2002.1011397 98

Authors' Biographies

PAULO MAZZONCINI DE AZEVEDO-MARQUES

Paulo Mazzoncini de Azevedo-Marques is a full-time Associate Professor of Medical Physics and Biomedical Informatics with the Internal Medicine Department, University of São Paulo (USP), School of Medicine, in Ribeirão Preto, SP, Brazil. He received his B.Sc. and M.Sc. degrees in Electrical Engineering in 1986 and 1990, respectively, and his Ph.D. in Applied Physics in 1994, from USP. He has previously worked on medical imaging quality control; since 1996, his research has focused on medical image processing. He held a research associate position at the University of Chicago in 2001, where he worked on medical image processing for computer-aided diagnosis (CAD) and content-based image retrieval (CBIR), under the supervision of Professor Kunio Doi. His main subject interest areas are CAD, CBIR, and picture archival and communication systems (PACS).

RANGARAJ MANDAYAM RANGAYYAN

Rangaraj Mandayam Rangayyan is a Professor with the Department of Electrical and Computer Engineering, and an Adjunct Professor of Surgery and Radiology, at the University of Calgary, Calgary, Alberta, Canada. He received the Bachelor of Engineering degree in Electronics and Communication in 1976 from the University of Mysore at the People's Education Society College of Engineering, Mandya, Karnataka, India, and the Ph.D. degree in Electrical Engineering from the Indian Institute of Science, Bangalore, Karnataka, India, in 1980. His research interests are in the areas of digital signal and image processing, biomedical signal analysis, biomedical image analysis, and computer-aided diagnosis. He has published more than 150 papers in journals and 250 papers in proceedings of conferences. His research productivity was recognized with the 1997 and 2001 Research Excellence Awards of the Department of Electrical and Computer Engineering, the 1997 Research Award of the Faculty of Engineering, and by appointment as a "University Professor" in 2003, at the University of Calgary. He is the author of two textbooks: Biomedical Signal Analysis (IEEE/Wiley, 2002) and Biomedical Image Analysis (CRC, 2005). He has coauthored and coedited several other books, including one on Color Image Processing with Biomedical Applications (SPIE, 2011). He was recognized by the IEEE with the award of the Third Millennium Medal in 2000, and was elected as a Fellow of the IEEE in 2001, Fellow of the Engineering Institute of Canada in 2002, Fellow of the American Institute for Medical and Biological Engineering in 2003, Fellow of SPIE: the International Society for Optical Engineering in 2003, Fellow of the Society for Imaging Informatics in Medicine in 2007, Fellow of the Canadian Medical and Biological Engineering Society in 2007, and Fellow of the Canadian Academy of Engineering in 2009. He has been awarded the Killam Resident Fellowship thrice (1998, 2002, and 2007) in support of his book-writing projects.

Index

Printed in the United States
by Baker & Taylor Publisher Services